The Last Alchemist in Paris

the LAST ALCHEMIST in PARIS

& other curious tales from chemistry

LARS ÖHRSTRÖM

OXFORD

UNIVERSITY PRESS

OXFORD

UNIVERSITY PRESS

Great Clarendon Street, Oxford, OX2 6DP,
United Kingdom

Oxford University Press is a department of the University of Oxford.
It furthers the University's objective of excellence in research, scholarship,
and education by publishing worldwide. Oxford is a registered trade mark of
Oxford University Press in the UK and in certain other countries

First Edition published in 2013
Impression: 1

Published in the United States of America by Oxford University Press
198 Madison Avenue, New York, NY 10016, United States of America

British Library Cataloguing in Publication Data
Data available

Library of Congress Control Number: 2013940788

ISBN 978-0-19-966109-1

Printed in Great Britain by
Clays Ltd, St Ives plc

CONTENTS

CONTENTS

PREAMBLE

THE PERIODIC TABLE AND THE
DA VINCI CODE

If you want some action, please proceed directly to Chapter 1. If you want a short introduction to the Periodic Table, the herding of electrons, and an idea of what Dan Brown could make of it, start here.

The Periodic Table of the Elements can be an object of fear for students. You may have problems with the irregular conjugations of the French verbs, or a tendency to muddle the order of the Edwards, Richards, and Henrys in the line of English monarchs, but the 114 elements of the Periodic Table, their symbols, and their places in this highly irregular jumble of little boxes seems to be of a different order of complexity.

To the initiated and the aficionados, the Periodic Table is a source of endless enthrallment, and to the apprentice chemists learning it by heart, it is their baptism of fire. To more normal people it just represents the chemical landscape we all wander in, though the relations between the map and our reality is sometimes rather vague. The stories in this book will try to bridge this gap between map and reality through the recounting of the adventures, successes, and misfortunes of ordinary and extraordinary people from around the globe in their intentional or unintentional meetings with various chemical elements.

H																															He
Li	Be																									B	C	N	O	F	Ne
Na	Mg																									Al	Si	P	S	Cl	Ar
K	Ca															Sc	Ti	V	Cr	Mn	Fe	Co	Ni	Cu	Zn	Ga	Ge	As	Se	Br	Kr
Rb	Sr															Y	Zr	Nb	Mo	Tc	Ru	Rh	Pd	Ag	Cd	In	Sn	Sb	Te	I	Xe
Cs	Ba	La	Ce	Pr	Nd	Pm	Sm	Eu	Gd	Tb	Dy	Ho	Er	Tm	Yb	Lu	Hf	Ta	W	Re	Os	Ir	Pt	Au	Hg	Tl	Pb	Bi	Po	At	Rn
Fr	Ra	Ac	Th	Pa	U	Np	Pu	Am	Cm	Bk	Cf	Es	Fm	Md	No	Lr	Rf	Db	Sg	Bh	Hs	Mt	Ds	Rg	Cn		Fl		Lv		

FIGURE 1 The 2012 Periodic Table according to the International Union of Pure and Applied Chemistry.[1] This is the long version that has the elements La–Yb and Ac–No in their proper places, and not broken out and placed underneath the other elements.

But before we start, I need to give you a rough guide to the geography and the map. In Figure 1 you see the Periodic Table in its 2012 version, written out in the so called 'long form', emphasizing the proper place of heavier elements such as uranium (U), and gadolinium (Gd)—akin to having the Orkney and Shetland Islands in their accurate positions on the map relative to the mainland UK, and not appearing in the oil-fields east of Aberdeen and Dundee. Or showing Alaska and Hawaii on the same grid as the mainland United States and not making them appear just south of California and west of Texas.

To give you some idea of why we draw it this way let me take you to an imaginary game park on the African savannah where two types of zebras, black striped and white striped, are idly grazing. There is only one water hole on the whole savannah, so all zebras need to go there at least once a day. The trouble is that these zebras are very aggressive. A white striped zebra will tolerate one, but just one, black striped zebra and vice versa; more than that and vicious fighting will result.

With only two zebras, one of each kind, there will be no problem. They will graze idly, avoid each other as much as possible, and a circular pattern will develop in the grass with the water hole

FIGURE 2 Fencing of zebras to keep the two types apart while still allowing them to reach the central water hole.

in the centre. If we want to have more zebras, and we probably do as these are beautiful animals, they need to be managed somehow to avoid fighting, and we plan to fence off the savannah in cake slices. However, the game park management will allow us to do this only in three different ways, where we can host a total of six, ten, and 14 zebras altogether in each set of pens (see Figure 2).

Electrons behave a bit like these zebras: they are attracted by the positive nucleus but want to avoid each other at all cost, as like charges repel each other. They can tolerate one neighbour, and this only if it has opposite 'spin', a property responsible for the everyday phenomena of magnetism but that is nevertheless hard to pin down. There are 'spin up' and 'spin down' electrons, but they are as difficult to distinguish as telling if a zebra has black or white stripes. Nature keeps them apart by fencing them in with a maximum of two, six, ten, or 14 electrons in each set of pens (which you now have to imagine as three-dimensional sections of space around the nucleus), only we call them *orbitals* and use the letters *s*, *p*, *d*, and *f* when we talk about them.

As we add protons to the nuclei to make heavier elements we also add electrons, and these have to end up in either the *s*, *p*, *d*, or *f* enclosures, and once we've started to fill a pen we will continue until

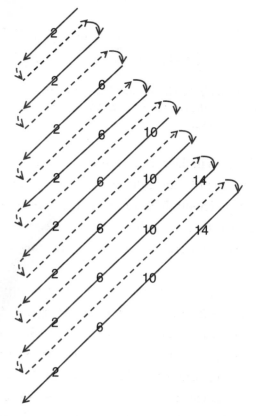

FIGURE 3 Finding a pattern for fencing in the electrons: each pen, or orbital, will take a maximum of 2, 6, 10, or 14 electrons. The line intersecting the numbers to the right traces a path through the Periodic Table.

it is full. The question is how many and in which order are they filled? Let's scribble the numbers on a piece of paper in a Dan Brown-ish way to form a pattern, as seen in Figure 3.

Then we draw a diagonal zigzag line through the numbers, and this will trace a walk through the Periodic Table in order of increasing atomic number: that is H, He, Li, Be, B, and so on. If we then replace, as we go, each element symbol with the maximum

```
[2]                                                                          [2]
[2][2]                                                          [6][6][6][6][6][6]
[2][2]                            [10][10][10][10][10][10][10][10][10][10][6][6][6][6][6][6]
[2][2]                            [10][10][10][10][10][10][10][10][10][10][6][6][6][6][6][6]
[2][2][14][14][14][14][14][14][14][14][14][14][14][14][14][14][10][10][10][10][10][10][10][10][10][10][6][6][6][6][6][6]
[2][2][14][14][14][14][14][14][14][14][14][14][14][14][14][14][10][10][10][10][10][10][10][10][10][10][6]     [6]
```

FIGURE 4 Periodic Table showing the four 'continents', those of maximum 2, 6, 10, or 14 electrons in the last-filled orbital. We also call them the *s*-, *p*-, *d*-, and *f*-elements, respectively.

number of electrons in the particular orbital we are just filling, we will get the picture in Figure 4. It is now evident, I hope, that the geography of the Periodic Table can roughly be described as having four 'continents', those of the 2, 6, 10, or 14 electrons in the last-filled type of enclosure, or as we prefer to say, those of the *s*-, *p*-, *d*-, and *f*-elements.

Chemists would not care for this numerology exercise if the map it produces did not help them manoeuvre among the atoms and molecules in the real world, but indeed it does.* Distinguishing elements by the type of orbital last filled, *s*, *p*, *d*, or *f* is very helpful, and a first step in understanding the chemistry of the elements.

Now and then chemists rearrange the Periodic Table into spiral form, three-dimensional edifices, circular or cylindrical versions. Periodic Table aficionados claim that there are at least 700 variations,[2] and some of these nicely enhance different relations between elements not apparent in the normal version. Others, however, it is argued, try to find underlying patterns or meanings in what is at the end of the day nothing more than a convenient way of presenting a large collection of data.[3]

---

* Quantum mechanics will, although it is a non-trivial exercise, even for physicists, come up with the same periodic table and many nice explanations as to the behaviour of the elements.

One could imagine Dan Brown having his 'symbology' professor-hero Robert Langdon[4] racing through a novel in quest of the 'true' representation of the Periodic Table, a feature that, if found, will prompt the world to 'disappear and be replaced by something even more bizarre and inexplicable'.[5] Or perhaps the answer to the ultimate question of Life, The Universe, and Everything in *The Hitchhikers Guide to the Galaxy* is not 42 but molybdenum?

1

Mr Khama is Coming to Dinner

If you know your Periodic Table you would perhaps have expected the first chapter to be about hydrogen, the lightest chemical element with atomic number one, consisting of only one proton in the nucleus with charge plus one, and one negative electron orbiting around the nucleus. However, this book will follow its own logic and we will start instead with the element once thought to be the heaviest in the universe, with atomic number 92.

The temperature is approaching +38°C, and the highway between Botswana's capital Gaborone and Francistown stretches out ahead in a straight line, heading north-east. It is the peak of the hot season, and here on the verge of the Kalahari Desert a dusty yellowish hue should colour the landscape, but instead rains have made everything a vibrant green. We stop and see hundreds of identical butterflies assembled in a small mud pool, and back on the well-paved and smooth road we are constantly vigilant to avoid the occasional cow, goat, or donkey feasting on the green grass beside the highway.

At the big coal mine and power station in Palapye we turn left from the main road, and after another hour we pass a big modern shopping centre and then, without really noticing, we have entered Serowe, considered by some to be the largest traditional village in Africa[6]—a settlement with a population of 90,000 or so spread out in mostly one- or two-storey houses in a very distinctive un-city-like manner.

We see signs directing us to the museum, that we don't find, and the *kgotla*, that we do. This large, very tidy, open space, surrounded by majestic trees and a wall at hip-height, is still the meeting place of the *Bamangwato* tribal councils (the word *kgotla* means 'court' in Setswana), but today it is completely deserted.

But let's now move back to 23 June 1949, when the situation was very different. Serowe, then the largest urban centre in the British Bechuanaland Protectorate, had just seen a massive invasion of South African and British journalists, in addition to the hundreds of tribesmen gathered in the *kgotla*. It was not as hot, as it was winter, but at that time this really was a remote place. There was not a metre of paved road in the Protectorate, the country was poor and austere, and the British preferred to conduct their fairly relaxed administration from the more comfortable Mafeking in the Union of South Africa.

The centre of attention on this day is a tall and fit man in his late twenties who is addressing the crowd, many of whom have journeyed a long way, and an absent woman. Who are they, and what exactly is at stake here?

The young man is Seretse Khama, the heir to the chieftainship of the *Bamangwato*. He is fighting for the recognition from his tribe of his marriage to Ruth Williams. Ruth is a young English woman made of stern stuff, formerly of the Women's Auxiliary Air Force and, because of this marriage, also an ex-clerk at a Lloyd's

FIGURE 5 Seretse Khama addressing the tribal court in Serowe, 1949. Photo © Time & Life Pictures/Getty Images.

underwriters in London. She was fired as soon as the marriage became public knowledge.[7, 8, 9]

Seretse is probably the only man in the country with a higher education, having a degree from Fort Hare University in South Africa as well as studies in Oxford behind him; currently he is reading law at the Inner Temple in London. The son of the former chief, he is designated by his uncle Tshekedi, the regent, to lead his people into the modern world, but falling in love and marrying a white woman was not part of the plan. Tshekedi and the elders of the tribe disapprove and demand a divorce.

But Seretse gains the support of the youngers of the tribe, and the mood of the delegates has swung in his favour. It could have ended here, with the *kgotla's* decision to welcome Ruth as their future queen, but instead this will turn into the 'Seretse Affair', a diplomatic and public relations nightmare for consecutive British cabinets, Labour and Tory, lasting well into the mid-1950s.

Why? According to the laws of the Protectorate, the British government had to confirm the new chief, and this was never to happen. First, the case was wrapped in layers of red tape: an enquiry was set up, the findings of the committee suppressed, copies of the report destroyed, and finally, in 1952, the couple was condemned without trial to permanent exile from Seretse's homeland by Lord Salisbury, Tory Secretary of Commonwealth Relations.[7, 8, 9]

While we cannot know with certainty which factors were decisive in influencing the actions of cabinet ministers and advice from senior officials, there is a remarkable coincidence that may indicate what might have tipped the scale.

At the beginning of the 1930s, atomic physics and nuclear chemistry were seen as little more than an expensive hobby for over-intelligent boffins, and countries emphasizing the usefulness of science, such as the Soviet Union, did little to fund such research. Consequently, radioactive materials were not in high demand. Radium was the most sought after, but even that did not amount to much, and the uranium ores that were its source had no uses except for colouring glass (which it does very nicely but, for obvious reasons, not any more).

As we all know, the situation changed quickly, with World War II, the Manhattan Project, the Bomb, and later the development of civil nuclear power, all relying on uranium. Even though this metal is fairly abundant (2.3 parts per million in the Earth's crust) and more common than tin, for example, workable deposits were few when demand set in. Moreover, the process from ore to metal was relatively complicated and unexplored, so new uranium mining and production could not be set up overnight.[10]

In 1939 uranium sat in a very remote corner of the periodic table. With an atomic number of 92 it was the heaviest known element, until 1940 when neptunium and plutonium were discovered, and only a select few knew about those two elements until after the war. In fact, its periodic geography was not quite settled. In 1939 it had not yet moved from its original home, placed right under tungsten (W).* The whole series of *transuranium* elements were still to be discovered and placed into their special category—the *actinoids*, with the atomic numbers 89 to 103—in the late 1940s.

During World War II, the US had obtained a virtual monopoly over uranium, controlling the two principal sources in 1949: Eldorado in Canada, and Shinkolobwe in the Katanga province of what was then the Belgian Congo.[11] The Soviet Union had to make do (or so it was thought) with the captured supplies remaining from the German atomic bomb project, and what could still be produced from the old Joachimsthal (Jáchymov) mine in the present day Czech Republic.

Although cheap and abundant atomic energy was no doubt an alluring prospect for the British government, national security in the form of nuclear weapons was probably higher on the agenda. British scientists had participated in the Manhattan Project, but even so the US did not share all its results with its former ally, leaving the Britons to work out the missing pieces for themselves, and, equally importantly, to find their own uranium.

---

* And in my 1946 edition of Niels Bjerrum's *Laerebog i uorganisk kemi* it still sits there, even though Bjerrum was Niels Bohr's old chemistry teacher.

When asked by nervous politicians, geologists predicted (quite accurately as it turned out) that in due time prospecting would locate sufficient uranium deposits to enable both long-term use of nuclear power as well as the development of atomic weapons. How they did that prediction is beyond the scope of this book, but geological maps showing the composition of the ground in terms of different rock types were already widespread in 1948, and given the type of rock, predictions could be made about which minerals were likely to be found in that location.

However, to actually find any uranium ore one needs to be out in the field, and with uranium there was help that could turn even the most amateurish stone collector into a uranium prospector*: the Geiger counter. This affordable handheld instrument measures radioactivity, although alpha-particles (helium nuclei with two protons and two neutrons ejected at high speed) are more difficult to detect than beta-particles (electrons) or gamma rays (similar to X-rays but of even higher energy). These three types of radiation are usually just called 'ionizing radiation', as they can strip electrons from nuclei to create charged ions; and that is also how they are detected by the Geiger counter.

When the radiation is passed through a tube filled with gas— for example, the noble gas neon—these gas molecules (or rather atoms, as neon is a one-atom molecule) will be hit by the high-speed particles or high-energy photons and electrons knocked out of the neon atom, giving a positively charged neon ion. Now the tube contains ions and can suddenly conduct electricity, just like a water solution of a salt (which neutral molecules cannot), and this is what creates the reading on the meter. But usually the counter

---

* As illustrated on the cover of US magazine *Popular Mechanics* in 1949.

also makes a very distinct clicking sound, to help the prospector home in on the source of the radiation easily.

Whether any of these amateurs actually hit gold and collected, at least in the US, a hefty reward from the government, I don't know. At the end of the war, however, for a nation wanting to acquire nuclear arms quickly, promises from geologists and prospectors of future mines were a small consolation. A convenient solution seemed at hand when it looked likely that low-grade ores from the Rand gold fields not far from Johannesburg, in the Union of South Africa, could be used to obtain uranium.[12,13] The pro-British Prime Minister of the Union, Field Marshal Jan Smuts, was eager to cooperate, and negotiations were already ongoing when his party suffered a landslide election loss in 1948, and the Nationalist Party under D. F. Malan came to power.

The new regime immediately started to implement its apartheid ideology, and in June 1949, coinciding with the Serowe *kgotla*, the Mixed Marriages Act was voted in by parliament without resistance. Consequently, a high-profile black-and-white married couple right on their doorstep would not be tolerated by the Nationalist Party. This was also the message from D. F. Malan to the British government after Seretse's triumph in Serowe. The exact consequences if the British did not cooperate were not spelled out, however.

That they had been under pressure from South Africa was long denied by British governments, who used a number of half-truths and groundless accusations as excuses for their actions.[7] However, evidence to the contrary was presented by the Cambridge historian Ronald Hyam in 1986,[14] and independently by journalist Michael Dutfield in his 1990 book *A Marriage of Inconvenience*.[9]

The Nationalist Party probably spent some time deciding what their best instrument of persuasion was, and it seems they chose uranium. Shortly after having delivered, in person, the first angry message from the South African Prime Minister, the Union's London High Commissioner Leif Egeland sent a note to the Secretary of Commonwealth Relations stating that the uranium negotiations would be suspended, at least until the end of October 1949.

Then, a third actor entered the scene, possibly pushing the uranium issue to the top of the agenda. On 29 August 1949, the Soviet Union surprised the world, and especially Western intelligence, by detonating its first atomic bomb more than three years ahead of the CIA's estimates.

The long and the short of the story is that Ruth, Seretse, and their baby daughter Jacqueline were exiled in England by two consecutive UK governments, notwithstanding the cabinets being under heavy fire from national and international press, and, it has to be said, with sincerely bad consciences among some (but not all) of the British officials involved. Prime Minister Attlee noted: 'It is as if we had been obliged to agree to Edward VIII's abdication so as not to annoy the Irish Free State and the United States of America.'[15]

In 1952, the first South African uranium plant became operative,[12, 13] and on 26 September 1957 the Khamas were allowed back home, although Seretse was never officially recognized as chief of the Bamangwato. By then there was, as predicted, plenty of uranium on the market, and South Africa was a lost cause for the Commonwealth anyway—Harold Macmillan's famous 'Winds of Change' speech in Cape Town being only two-and-a-half years away.

The uranium story stops here, but not the Seretse and Ruth Khama story. Readers of Alexander McCall Smith's delightful books about *The Number One Ladies' Detective Agency* might have noted a picture hanging on Mma Ramotswe's wall: that of

Botswana's first President, Sir Seretse Khama, 1921–1980.* Mma Ramotswe holds him great esteem, ranking him equal to the Queen and Nelson Mandela.[16]

So, was the Khamas' forced exile the consequence of South African blackmail over a uranium contract? We do not know for sure. Ronald Hyam and Peter Henshaw argue in *The Lion and the Springbok: Britain and South Africa Since the Boer War* (2003)[17] that the UK government was more concerned with the threat of direct annexation of its southern African protectorates by the Union, and that the Khamas' exile was seen as a small price to pay in order to protect the inhabitants of present-day Botswana, Lesotho, and Swaziland from falling under the yoke of apartheid.

At the same time, Hyam and Henshaw note that for most cabinet members this was a question of a strategic nature, including access to important raw materials, based on 'the context and imperatives of the Cold War'. The vulnerability of the protectorates was of most concern to the ministers and their civil servants. There seems, however, to be no evidence in the British archives for a direct link between the suspension of uranium negotiations in 1949 and any actions taken against the Khamas. The answer to whether or not the South Africans were actually playing the uranium card in this game needs to be researched in the archives of Pretoria.

My personal view is that for a short time in 1949, following the detonation of the Soviet Union's bomb, the uranium question may have been important, but that for the overall actions taken by the UK government in subsequent years it was only one of several secondary factors influencing their decisions—racial prejudice being another.

* Lady Ruth Williams Khama, 1923–2002.

There are those who say that the Ruth and Seretse story was one source of inspiration for Spencer Tracy's last movie, *Guess Who's Coming to Dinner*, also starring Sidney Poitier, Katharine Houghton, and Katharine Hepburn, and directed by Stanley Kramer. In this classic Hollywood production,* Houghton—a young white middle-class woman—invites her, very recent, fiancé Poitier, who happens to be black, to dinner with her parents. The film was released in December 1967, six months after the United States Supreme Court outlawed the banning of inter-racial marriages. At that time these laws were very much in use in 17 states of the union, and the 'crime' punishable by jail. The last state to officially remove its so-called anti-miscegenation act from the law books was Alabama, in the year 2000.[18]

Why was the CIA estimate for the Russian atomic bomb so embarrassingly off the mark? In Chapter 2 we explore the part of the Periodic Table that provides the answer.

---

* The film received two 'Oscars' (Academy Awards): Hepburn for Best Actress and William Rose for Best Original Screenplay.

2

From Bitterfeld with Love

In Chapter 2 we enter the deceptive and shadowy world of espionage or 'intelligence', and in the process we will learn how to make metal out of rocks.

In September 1961 Henry Lowenhaupt threw his last pieces of East German calcium metal into the Potomac River, and watched the violent reaction that made the water boil as the metal released two of its electrons to the H_2O molecules, producing hydrogen gas and large amounts of heat in the process. Spelling out chemical reactions may seem scary, but it is really quite simple and I will not hesitate to show one or two equations as we proceed. Just remember that no atoms or electrons ever disappear and you will be fine.

What I just stated above would translate into:

$$Ca + 2H_2O \rightarrow Ca^{2+} + H_2 + 2OH^-$$

This may look somewhat similar to the reaction of sodium metal with water that many of us has seen demonstrated in school, and that is certainly true. The calcium reaction is a bit slower however, and somewhat less energy is released, because two electrons are removed from the atom, not just one as for sodium.* (Removing electrons is what we call an oxidation, and calcium has changed from oxidation state zero to oxidation state, or number, +II.)[†]

Mr Lowenhaupt would have known all this. He was a Yale University graduate who had worked for the Manhattan Project making the first atomic bomb, and spent the rest of his career with the CIA, from its inception in 1947 until his retirement in 1991.[19] The calcium metal that went into the Potomac was the last remains of a project that could have saved the CIA from its first major embarrassment, and that was part of one of the most sophisticated sabotage operations ever conceived: *Operation Spanner*.[20]

As we saw in Chapter 1, the detonation of the first Soviet atomic bomb in 1949 came as a surprise to both the CIA and MI6. But to their credit—for the Americans at least—one should perhaps point out that it also came as a surprise to the Russians that they were immediately found out. (The setup of this detection system is in itself quite an interesting story.) Why, then, did the US and UK intelligence services underestimate the momentum of Stalin's atomic bomb programme so drastically? In essence, they misjudged the ability of the Soviet Union to produce uranium. Low-grade ores from mines in the Ural Mountains were used, in addition to existing supplies captured from the Germans.

While spying on the actual Russian atomic energy sites was more or less out of the question, much information could be extracted from other sources. One principal target was both the personnel and the installations used by Nazi Germany's *Uranverein*, the

---

* With sodium metal this reaction generates so much heat that the hydrogen gas is sometimes ignited.
† We will normally refer to oxidation states with Roman numerals.

Uranium Club, the codename for the nuclear energy and weapons programme that had come under Russian administration in their occupied zone. One key issue was how the Russians were going to make uranium metal out of the uranium ore.

The general rule is that metallic elements are unstable in their metallic, neutral form (oxidation states zero), with notable exceptions such as the noble metals gold, silver, and copper, which you can find as nuggets if you are lucky. These metals hold on very tightly to their electrons, in stark contrast to sodium or calcium, which are just waiting for something suitable on which to dump the electrons they have in their last filled 'pen', or 'orbital'.*

Uranium may be as good as gold in some circumstances, but it is certainly no noble metal. It is usually found in nature with oxidation number IV, U^{4+}, with four electrons removed, normally combined with oxygen into UO_2 (or into the mineral *pitchblende*, with the approximate formula U_3O_8 that has a combination of U^{4+} and U^{6+} ions). To put the electrons back and get uranium metal we need something that is very willing to give away its electrons— what we call a strong reducing agent.

Most of the chemical reagents used in producing uranium from uranium ore are ordinary and have other uses, but because uranium is so ignoble, it needs a very special and strong reducing agent to become metallic. The Germans had used calcium metal produced in Bitterfeld, a small town in eastern Germany close to Leipzig and not far from the old uranium mine in Joachimsthal (Jáchymov) in the Czech Republic. Because Germany was divided by the occupying powers, and the German Democratic Republic

---

* We called them this in the Preamble; 'shell' is the term you may have encountered in school.

with its strict borders controls was yet to be established, getting 'intelligence' from inside the Russian zone was at least feasible, and both MI6 and CIA kept a close watch on the Bitterfeld factory (and on each other).

In 1947, evidence had been gathered to show that the Bitterfeld factory was producing 30 tons of high purity distilled calcium metal every month. We associate distilling with the separation of ethanol from water—making spirits at distilleries. The method is based on the different boiling points of the two substances, 78°C for ethanol and 100°C for water, but distillation is in fact a very general method of purification in the chemical industry, applicable to all sorts of substances. If you cool air enough so that it becomes liquid (it has to be very, very cold) you can distil it and separate out the components, such as nitrogen or neon.* Calcium melts at 842°C and will boil at 1484°C, but these temperatures are lowered if the pressure is reduced. However, you don't need to build your chemical plant on the top of the Himalayas to achieve this, as it is quite easy to create similar conditions in the factory. Indeed, in 1946 the US had already put vacuum pumps on the 'export control list', stopping a major order by a Russian trading organization.

The question facing Lowenhaupt and his colleagues was that they needed to be sure that the 30 tons of calcium per month was going to the Russian nuclear programme, and not to some legitimate use in German industry. After some probing they established that about 5 tons of rather impure metal per month had been produced for the Osram and Philips companies during the war,

---

* This is a major industrial process. Neon and the other noble gases have various uses, and liquid nitrogen is important for cooling but needs to be separated from liquid oxygen, which is a very dangerous substance.

apparently for making radio-tubes, and a further 20 tons per month of different calcium alloys (a mixture of metals) with aluminium and zinc was sold to the German railroads. There was thus no way that the 30 tons of distilled calcium had any buyers in German industry, and when an agent inside the Bitterfeld factory reported that three rail cars carrying distilled calcium had left on 26 July 1947 with the destination 'Elektrostahl Moskau' Post Box 3, Kursk Railroad, it should have been clear that the cat was out of the bag.

To further investigate what the Russians planned to do with the calcium, a sample was smuggled out by the agent inside the Bitterfeld factory, and when the CIA obtained the complete analysis it was clear that it had all the specifications needed for reducing uranium ions to high purity uranium metal ready for nuclear applications. The way to achieve this is to first make UF_4 molecules, and then to react these with the calcium metal to produce calcium fluoride and uranium metal, more conveniently written as:

$$2Ca + UF_4 \rightarrow 2CaF_2 + U$$

You don't explicitly see the electrons changing places in this reaction, but as the fluorine is always considered to be charged minus one, except in F_2 gas, it is quite easy to work out that uranium starts out as +4 and the two calcium atoms end up as +2 each.

The nice thing with these reaction formulae is that we can now calculate exactly how much uranium metal the Russians could make each month. I will not bore you with the details,* but it is

---

* But here they are nevertheless. For every uranium atom you need two calcium atoms. As the atomic weight of uranium is 238 and that of calcium 40 this means that for 80 grams of Ca you can get 238 grams of U, and 30 tons of Ca will give you, at least in theory assuming the reaction goes really well, 30*238/80 = 187 tons of uranium.

really not much different from calculating how much mayonnaise and how many meringues you can make from a certain number of eggs. As uranium is much 'heavier' than calcium you can, in theory, get almost 200 tons from 30 tons of calcium. In real-world processing this number is lower, and a large excess of calcium is used. The Americans calculated using a technical limit of about 1:2.2, such that 30 tons of calcium would only give 66 tons of uranium.

Still, estimates from these figures indicated that the Russians had much more uranium than that predicted from the known uranium sources available to them. For reasons that are unclear—Lowenhaupt blamed 'hubris'—this information was not properly acted upon; otherwise the date for a first Russian bomb would surely had been revised. Whether that would actually have made any difference to anything is not clear, but what could have had an impact was *Operation Spanner*, devised by legendary MI6 officer, and chemist, Eric Welsh.

The most important specification for the calcium metal was the level of boron impurities. Boron, symbol B with atomic number five, naturally exists in two forms that we call *isotopes*, one with five protons and five neutrons in the nucleus, ^{10}B, and one with an extra neutron, ^{11}B, or boron-11 (the use of super and subscripts is demonstrated in Figure 6). For the nuclear scientist boron-10 was the big problem, as this atom would gobble up any neutrons fired at the sample to split uranium atoms (and secondary neutrons from this split that made plutonium) and become boron-11. With too much boron-10 in the uranium metal the nuclear reaction would simply stop.

Normal chemical analysis of the calcium metal leaving the factory in Bitterfeld would not reveal how much of each isotope was

3

The Curious Incident of the Dog in the Airship

In Chapter 3 we play with gases and fire, and explore one of the most important equations in science.

Joseph Späh had to feed his dog; nothing strange about that. The problem was that Ulla, an Alsatian, was mostly confined to the freight room—off limits to passengers. Had everything gone according to schedule, this would not have been an issue either, except for gruff remarks from crewmen not appreciating the needs of this canine friend and co-worker in Späh's stage act.[23] But this flight did not go according to plan, and Späh's frequent visits to the rear of the *Hindenburg* would give him problems in the years to come.[24]

The US Department of Energy, and its counterparts in Europe and Japan, are currently spending billions on developing the use of hydrogen for future energy applications—for example, as a fuel for cars and buses.[25] The main advantage is the clean combustion of this fuel: two molecules of hydrogen gas will combine with one molecule of oxygen and give two molecules of water. The future belongs, perhaps, to the 'hydrogen economy', but unfortunately for its proponents, the popular history of hydrogen as a fuel is bound up with the tragedy of the *Hindenburg*.

We will get back to Joseph Späh's poor dog in a while, but for now ponder the fact that over the dog, and above everyone else aboard the comfortable and luxurious *Hindenburg*, there were huge 'bags' filled with hydrogen—the lightest of all the elements, with only one proton and one electron. It has the lowest density of any gas, and is formed by two hydrogen atoms combined together via a single chemical bond, made by sharing the two negatively charged electrons between the two positively charged nuclei. This H_2 gas had carried the world's largest airship from Frankfurt to Lakehurst outside New York, and before that on successful tours all over the globe during the preceding year.

These days, we tend to wonder how people could even contemplate the idea of travelling around in what can be described as a flying bomb. But then we forget that the German commercial airships, manufactured by Luftschiffbau Zeppelin GmbH, had an excellent safety record.[26] The *Graf Zeppelin* for example completed eight years of regular service—chiefly on the Germany–Brazil route—without incident, and the big airships worldwide, with few exceptions, all used hydrogen gas as their lifting power and had done so for about 30 years.

So, in the asbestos-clad smoking lounge aboard the *Hindenburg*, Späh and his fellow passengers leisurely enjoyed whatever view there might have been on the Atlantic crossing on this the first regular Frankfurt–New York service in 1937, and thought nothing of the hydrogen. A home movie made by Späh, miraculously surviving the final disaster, shows them enthusiastically pointing at some icebergs floating by as the airship approached the American continent at the dignified speed of 134 kph.[27]

How stupid, we think nowadays: why didn't they use helium instead? And then we remember—there was a US trade embargo against Nazi Germany, so the Germans could not buy any helium. But when you check the details it turns out that the story is both more complicated and more interesting than this simple and, as we will see, incorrect but well-known 'fact'.[28]

The engineers of course knew all about helium. With two protons, two neutrons and two electrons this monoatomic gas is denser than hydrogen, and so does not have quite as much lifting power. We can deduce this, since both H_2 and He (and almost any other gas we can think of) obey a wonderful law of chemistry called the *ideal gas law*. Among many other things, the equation of the ideal gas law tells us that if we double the weight (or more correctly, the mass) of the gas molecule the density of the gas will be doubled too.

However, there were also technical advantages to helium. Being twice as heavy per molecule means that helium is a better insulator than hydrogen, a factor not to be ignored for an airship baking in the sun all day. The heating problem comes from the fact that a gas obeying the ideal gas law that increases its temperature 10 per cent (expressed in degrees Kelvin) will also increase its volume by 10 per cent (or, if it is enclosed in a fixed volume container, its pressure by the same proportion). Both effects had to be carefully calculated by the engineers to make sure that both the gas bags and the airship's aluminium framework could withstand such mechanical strains. Why is this then a lesser problem with helium? Because helium heats up more slowly, as the average speed of the gas molecules is lower if they are heavier, and this means that helium transmits heat by conduction at a slower rate. This we can also derive from the ideal gas law (maybe you can see why it is so wonderful!), and in the colder parts of the world we put this

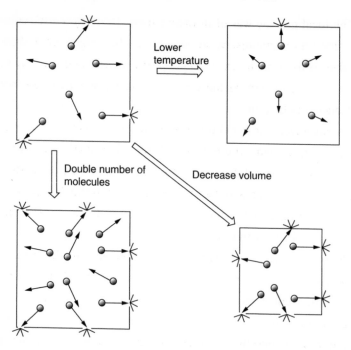

FIGURE 7 The Gas Law tells us how a gas behaves under normal conditions. Upper left: pressure (P) on the walls of the gas container, for example a helium balloon bought at the zoo, arises because gas molecules hit the wall, and the more hits per second the higher the pressure. Upper right: when the molecules are cold (low T) and move slowly, fewer of them will hit the walls every second and the pressure will be lower too. Lower left: if we increase the number of gas molecules (n), more will hit the walls every second. Lower right: if we decrease the volume (V) there is less space to move around in and again more molecules will hit the walls. All elegantly summarized in the formula $P \times V = n \times R \times T$,[*] where R is the 'gas constant', a number with a wider significance than just explaining the behaviour of helium balloons.

knowledge to good use in double- or triple-glazed windows filled with argon, a gas with molecules having a mass greater than nitrogen, the main constituent of air.

[*] n is number of molecules, usually expressed as number of moles, and T is the temperature in degrees Kelvin. (6×10^{23} molecules make 1 mole.)

The ideal gas law was of immense importance for early scientists developing chemistry, but also for airship engineers designing their ships to withstand changes in the volume and pressure of a gas, and also adapting the design to the different lifting gases hydrogen and helium.

The big advantage of helium, however, is that it is the least reactive element known—there is not a single known chemical compound containing helium atoms combined with other elements. Thus no dangerous reactions can take place with this gas, in contrast to hydrogen gas, which readily, and in some proportions explosively, reacts with oxygen.

The big difficulty for Germany, and indeed also for Britain and France, was that there was no helium to be found in Europe. The US was the world's only provider, and in the 1920s it was in scarce supply even there.

Helium is created by the radioactive decay of heavier elements in the Earth's interior, especially thorium and uranium. It seeps through leaks and cracks in the rocks and is recovered as a minor part in natural gas. As oil was the thing to drill for in the 1920s, natural gas production was low and so was helium production. It is said that when the US Navy launched the first of four gigantic helium-based airships, the *USS Shenandoah* (Figure 8), in 1923, its gas bags contained most of the helium ever produced, and that when *USS Los Angeles* was taken into service in 1924 it was difficult to operate the two airships at the same time due to helium shortage.

So helium was seen as a major strategic resource, its production was put under the supervision of the Bureau of Mines, and the exports were ultimately controlled by the US Secretary of the Interior. A large national reserve (still in use in 2013) was created

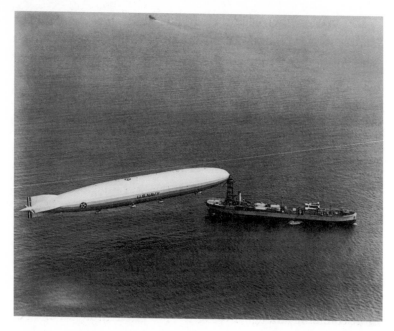

FIGURE 8 If the dating of this photo to 1924 is correct, the airship in the picture, the *USS Shenandoah*, probably contains almost all commercially available helium existing at that time. © CORBIS

in 1925. Eventually though, production increases filled the stocks and export licences were subsequently granted.

Among the public, however, the notion of a US helium export ban seems to have lingered, as President Herbert Hoover felt the need to comment on the issue during a press conference in 1930, saying 'So that is an entirely mistaken notion that the United States is preventing the use of helium in the development of lighter-than-air navigation'. He then went on to explain that the reason the helium export was not taking off was that it was four times more expensive than hydrogen, and that hydrogen could be 'made on the spot', whereas there was a considerable lack of helium 'service stations' around the world.[29]

You may wonder where we get hydrogen from, as President Hoover thought that its easy availability all over the planet was self-evident. Part of the answer is that there are many ways, and being a mining engineer himself he probably knew this very well.

In the prelude to another famous disaster—the 1897 Swedish North Pole expedition of Andrée, Fraenkel, and Strindberg (a relative of the author of the same name whom we will meet in Chapter 12)—pig iron and sulphuric acid were carried on the steamer *Virgo* to the Svalbard Islands where the acid was poured on the iron. The hydrogen gas formed in the reaction was collected and used to inflate the balloon *Örnen (The Eagle)*, which soon lifted and drifted away to the north, never to be seen again.

$$H_2SO_4 + Fe \rightarrow Fe^{2+} + SO_4^{2-} + H_2$$

More commonly though, hydrogen is produced through a number of clever and related processes using carbon and water—with natural gas or oil as starting materials*—but it can also be made by the direct electrolysis of water using an electric current: a reaction that is, in fact, the reverse of the potentially explosive combustion reaction.

So, because of this ease of production and the abundant possibilities of emergency 'refuelling', hydrogen continued to be the gas of choice for German and British airships, and the statement from Hoover (dated 10 October 1930) might well have been a direct response to the British *R101* accident.

The R101 was one of the two first ships in the planned Imperial Airship Scheme, and it crashed on its maiden voyage outside

---

* This, by the way, is not very sustainable. What we really want to do today is to produce hydrogen with the help of sunlight, either through some clever photocatalytic process or via solar cells and electrolysis.

Beauvais in northern France, a few hours after crossing the Channel. It was a government ship that carried British officials on their
way to Karachi (in what is now Pakistan), but hopes were high
that the *R101*, and the Vickers-constructed competitor the *R100*,
could both become commercial successes.[30]

The attractiveness of the airships can be seen in comparing
schedules on this route: the Imperial Airways service took eight
days, with 21 stops on the way—by sea the journey took four
weeks. The *R101* promised to do the same in five days with only
one stop (Ismalia in Egypt), and what was more, with the comfort
of 'a lavish floating hotel'.

The crash itself (in the early hours of 5 October 1930, five days
before Hoover's press conference) was not really violent, and in
different circumstances would only have caused minor injuries. The
following fire, however, killed all but five of the 49 crew and passengers. Obviously, the burning hydrogen gas was the destructive
agent here, but the underlying causes were a number of fatal admin-
istrative, political, and technical decisions made by Lord Thomson,
the Secretary of Air, and other senior officials, most of whom lost
their lives in the accident. At least this is how it is told 25 years later
in British–Australian popular author Nevil Shute's autobiographical
book *Slide Rule*.[30] At the time, Shute was chief engineer of the competing Vickers *R100* airship, which had safely completed a return
test flight to Canada earlier the same year, and he had supreme
insights into all aspects of the airship programme. The accident put
a full stop to the Imperial Airship Scheme, and evidently fostered
Shute's life-long aversion for government officials and rather ideal-
istic views on private enterprise and entrepreneurship.

The *R101* disaster, and an increasing supply of helium available
from the US, probably caused the Zeppelin engineers to rethink

their designs, and when plans for the two sister ships, later to be named *Hindenburg* and *Graf Zeppelin II,* were drawn up in 1931, helium was considered for the lifting power. However, two political changes were going to make this impossible: the coming to power of Adolf Hitler and the National Socialist party in Germany, and the election of Franklin D. Roosevelt as president of the United States.

Roosevelt himself was not the main obstacle, but he had chosen as his Secretary of the Interior, and thus commander of the national helium reserve, the fiercely independent and anti-fascist Chicago politician Harold Ickes.[31] Ickes would not sign any export licences for helium to Germany, and the *Hindenburg* thus had to be fitted for hydrogen. It is important to note, however, that there was no general US boycott or trade embargo with Nazi Germany.

Most people have at one time or another seen the footage of the *Hindenburg* accident from the Lakehurst airfield. Most people are probably also convinced, as I was, that nobody could have survived the crash and the flames. Everything happens extremely fast, and in seconds the whole ship is on fire. I had a nice surprise when I found out that this was not the case at all: more than half the passengers and crew were saved, and some, like Joseph Späh, sustained only minor injuries (many were of course badly wounded, and of the 97 on board 35 died, along with one member of the ground crew who was also killed).

It is because of a not-so-wonderful law of media that the Beauvais *R101* disaster is largely forgotten, although it can be argued that it was far worse, while the *Hindenburg* accident is known all over the world: if it is not on film, it's almost as if it didn't happen.

There seems to be no general agreement about the cause of the *Hindenburg* accident. The Encyclopaedia Britannica tells us: 'The fire was officially attributed to a discharge of atmospheric electricity in the vicinity of a hydrogen gas leak from the airship, though it was speculated that the dirigible was the victim of an anti-Nazi act of sabotage'. There has never been any evidence for foul play, but as the fire started in the rear of the *Hindenburg*, Späh's frequent visits to his dog, witnessed by many members of the crew, seemed suspicious. After the accident he was thoroughly investigated by the FBI and finally cleared of any suspicion.

The accident shocked the world, and new demands were made from Germany for helium to be used on the *Graf Zeppelin II*. Ickes had the whole cabinet against him, including President Roosevelt, and finally gave in.[32] A *New York Times* article from 17 January 1938 reported that the 3,663-ton German freighter *Dessau* was expected shortly at Houston to load the first helium shipment to Germany.[33] However, Nazi politics again jeopardized the helium agreement, as in February the Austrian chancellor was 'invited' by Hitler to make a deal that prepared the complete 'Anschluss' of Austria to Germany a month later.

In Ickes' diary, published after his death, the helium question pops up continually during the spring and early summer. For example, the first entry of 17 April is: 'I have not yet signed the helium contract', and he describes recurrent heated arguments with both the President and the Secretary of State, Cordell Hull, inside as well as outside formal cabinet meetings. He was even visited in person by the Zeppelin Company's Dr Captain Eckener, an outspoken anti-Nazi of whom Ickes had 'a very high opinion', but to no avail. Ickes persisted, and won—the contract was never signed.[34]

Joseph Späh's dog Ulla sadly did not make it out of the *Hindenburg*, but Späh continued to perform as an acrobat under the name 'Ben Dova' until the beginning of the 1970s, when he retired.

The *Hindenburg's* sister ship *Graf Zeppelin II* did play a minor role in Nazi propaganda in the 1930s, and flew a few spy missions in the advent of World War II, but was soon decommissioned and the aluminium frame ended up in Messerschmitt fighter aircraft. But I urge you to visit the wonderful world of the Internet and watch the film of the *Hindenburg* in all its glory, cruising over Manhattan in 1936, and dream of a more dignified way of travel.

But do not imagine yourself cruising in a helium lifted airship to your favourite holiday destination. Now, as in the 1920s, we have a helium shortage and what makes matters worse, helium is so light that it will escape the Earth's gravity field. This means every kid's ballon bought at the park is a valuable resource wasted for eternity.

4

The Spy and the Saracen's Secret

A chapter in which we learn not to be too nosy around Sheffield, and how to find a map that will help us make steel.

The fifteenth of August 1754 was not a good day for the Swedish spy Reinhold Angerstein. After allegedly showing a very keen interest in Benjamin Huntsman's Crucible Steel Works in Sheffield, he seems to have been promptly invited to take the next coach out of town, or something similar. In his diary he is only able to record some superficial information on how to make pocket-knives, and compared to the details given about other places, the lack of information about Sheffield is very suggestive of an early and speedy departure.[35]

However, a man of considerable social talents, as well as a gentleman, he did not need to despair. He apparently found lodgings for the night with the young Marquess of Rockingham, Charles Watson-Wentworth at Wentworth House. In passing, between technical and business-related notes in his diaries, we are told that the Marquess is married to 'a daughter of a rich gentleman'.[36] He does not let us know if she is pretty or not, but maybe old-time spies were more single-mindedly focused on their assignments. He had certainly not read Ian Fleming, and probably had no idea how we expect a gentleman-cum-spy to behave. We are, however, told that one of the Marquess's ancestors was beheaded for supporting Charles I.

He himself was apparently not bad looking, if the likeness of the portrait in Figure 9, still hanging in the head office of Jernkontoret ('The Iron Bureau', the Swedish Steel Producers' Association)* in central Stockholm is to be trusted. In 1754 he was 36 years old and, with some poetic licence, we could say that he was travelling in England and Wales in order to discover the Saracen's secret. He was an industrial spy sent out by the Swedish government, and as such he used 'all possible means, legal or otherwise', to get to see what he wanted.[35]

In the mid-eighteenth century, the Swedish government and the country's iron manufacturers wanted to know everything about British steelmaking. The reason is not that the Swedes wanted to improve their arms factories—any realistic dreams of Swedish military power in Europe ended in the small Ukrainian town of Poltava almost 50 years before. No, this was simply business. A large proportion of the iron worked in the British steel industry came from Sweden at this time, and accounted for as much as 60 per cent of total Swedish export revenues in certain years.[35] There was thus a good reason for the Swedes to keep an eye on any iron-related developments in England and its neighbours.

Now, I would of course like to go on to say that the reason for Angerstein's speedy retreat from Sheffield and its steelworks was

---

* Swedish 'jern' is the old spelling for 'järn' meaning iron.

FIGURE 9 Reinhold Angerstein, gentleman, industrialist, and spy, nowadays overseeing a fax and copying machine at the Swedish Steel Producers' Association in Stockholm. Portrait painted in 1755 by Olof Arenius, photo by the author.

that the owner, hawk-eyed Mr Huntsman, was guarding the Saracen's secret—an ancient steelmaking procedure he had just rediscovered—from all possible competitors.[37] That, however, would not quite be truthful, as what I choose to call 'the Saracen's Secret' is a pretty complicated affair and, despite claims to the contrary, not fully understood even today. Nevertheless, I would not be far off either, as this is all about making high-quality steel out of iron and carbon.

Pure metallic iron is not a very useful material. It rusts easily and is a relatively soft metal. But mix it with a little carbon and you get steel, a material that has physically changed our world in more ways than one cares to count: from skyscrapers and bridges, to the surgeon's blade and the gold miner's drill.* The discovery of steel is one of nature's mesmerizing coincidences. In order to make iron metal out of iron ore, early metallurgists used carbon-containing materials in the form of dry wood, and as processes were refined this was replaced by carbon in the form of charcoal, all leading to small amounts of carbon finding its way into the iron metal and making the magic mixture.

The carbon material not only provides the heat needed for the melting of the iron, it is also a crucial reaction ingredient. We learnt in Chapter 2 that metals in nature mostly exist as positive ions that need electrons for them to become metallic. Uranium needed the extraordinary reducing agent metallic calcium, but to make iron metal from Fe^{2+} or Fe^{3+} ions we can use carbon instead:

$$3C + 2Fe_2O_3 \rightarrow 3CO_2 + 4Fe$$

Here we work out the oxidation states, or numbers, by using the rule that oxygen is always charged −2, except in elemental form or combined with fluorine, and see that iron starts as +3 and carbon ends up as +4.

---

* Technically speaking, steels are those iron materials containing less than 2 per cent carbon by weight. Those with higher carbon concentrations have different names, for example cast iron.

The reaction above is a simplification: in a blast furnace many reactions take place simultaneously, and iron is mainly reduced by carbon monoxide, CO, formed when the large excess of carbon is partially oxidized by oxygen from the air. This reaction, which is also an approximation, you may have come across in school chemistry:

$$3CO + Fe_2O_3 \rightarrow 3CO_2 + 2Fe$$

When the iron melts it will dissolve some of the carbon, and as the carbon atoms are smaller than the iron atoms (about the same ratio as a billiard ball to a golf ball)* they will not completely destroy the atomic structure of the iron. Instead, when an iron–carbon melt cools and starts to solidify, the small carbon atoms will insert themselves in between the iron atoms, with the general effect of making the material harder but more brittle.

We can think about this in the following way. Without the carbon, the iron atoms have more freedom to move around making the material strong, as a force applied to the material can be met by miniscule movements of the atoms. When the empty spaces are occupied by carbon atoms there are many more atom–atom attractive interactions, some on the border of normal chemical bonds, making the material much harder but at the same time more brittle, as there are fewer ways of rearranging the atoms to withstand an external force. Or you could think of the pure metal as having the atoms embedded in a tough electron jelly: when we add carbon atoms, this jelly is partly replaced by ball-and-spokes-type bonds between the carbon and iron atoms—bonds that are strong, but once broken remain broken.

---

* The actual values for the atomic radii are 0.07 nanometres for carbon and 0.14 nanometres for iron.

Not only is the carbon content important, but the time spent at different temperatures, the rate of cooling, and the addition of other alloying metals are also crucial. This makes steel, old fashioned as it may seem, a high-tech material, as metallurgists and materials scientists continue to discover innovative ways of producing new varieties.

In the first 4,000 years or so of steel making, the early chemists and metallurgists had no real idea what they were doing, and thus found it very difficult to optimize their processes. Add to this difficulty the large and very diverse selection of iron ores found in nature—frequently with phosphorus and silicon atoms causing a nuisance—and you may appreciate some of the complexity of the problem. Simply copying a successful procedure might not give a satisfactory product with iron ore from another mine. During his travels through England, Angerstein duly took note of the origin of the raw materials used in the various iron works around the country, being no doubt pleased to notice that the best steel was made from ore from the Dannemora mine, north of Stockholm.

What the early steelmakers lacked was a good map and a way to look at the atomic details of their products. A simple form of such a map is displayed in Figure 10, more correctly called an iron–iron carbide phase diagram.

This map has two types of coordinates. If you travel from left to right you go from a material that is pure iron to one that contains one carbon atom for every four iron atoms (20 per cent carbon based on the number of atoms, not on the weight, otherwise written as '20 atom per cent C'). If you travel from the bottom to the top the temperature increases from 600°C to 1600°C, and eventually you get a liquid (or if you prefer, a melt)—the dark grey area. The areas immediately below the dark grey at the top are solid–liquid mixtures (like ice and water co-existing at 0°C), and below those we have distinctive solids (or steels/cast iron) differing in their detailed atomic structures.

In the left half of the diagram (under 9 atom per cent) you have steels, and further to the right cast iron. The *Kirk-Othmer Encyclopedia of Chemical Technology* takes two pages to explain this diagram to the initiated chemical engineer, so I will not go into any details but for two things. The quality of the steel depends crucially on which part of the diagram you are in, and by cooling fast enough you can, for example, make a steel that retains its high-temperature structure and never transforms into the form you should theoretically have in the low-temperature areas. You can also make steel with different atom arrangements on the surface and on the inside.

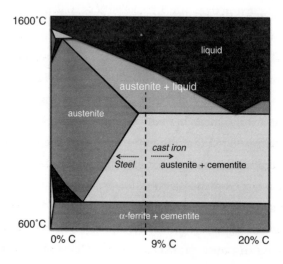

FIGURE 10 The iron–carbon steel map from 0–20 atom per cent carbon and from 600°C up to 1,600°C (below 600°C nothing will change), also known as the iron–iron carbide phase diagram. Over 9 atom per cent there is cast iron, below there is steel. The names indicate different phases with different atom arrangements. The diagram is a simplification.

So the secret Benjamin Huntsman had in Sheffield was a better way of controlling and optimizing the percentage of carbon in the iron, and their different atomic arrangements, making his steels the best available at the time.

However, many hundreds of years earlier, perhaps in present-day Syria, another particularly good combination of processes, craftsmanship, and raw materials had occurred, creating a legendary steel and a legendary weapon: the Damascus sword. Apart from its deadly properties, and the activity it led to as sword smiths all over Europe tried to copy it, the writer Walter Scott used it in the novel *The Talisman* as a metaphor to reveal the unsophisticated culture of the Europeans of the Third Crusade (1189–92) compared to the refined Muslim world.[38] In the novel, during a meeting between Richard the Lionheart (Cœur de Lion) and Saladin (Salahuddin Ayyubi), the English king demonstrates the power of his sword—cutting a bar of iron in two using great force. The Sultan of Egypt and Syria then draws his blue gleaming Damascus sword and, with a light hand, completely effortlessly cuts a soft cushion into two—a feat Richard had deemed impossible for any conceivable sword just moments before.

It has been alleged that for a long time Damascus swords and armour were superior weaponry, and that no European imitations ever quite matched them. The truth of this is, however, disputable.

The image of a superior shimmering blue sword with intricate patterns in the blade, forged by mysterious Eastern blacksmiths, and carrying an exotic name is of course irresistible to romantics like myself. Sir Walter Scott was no novice when it came to armour. His collection can still be seen in his home at Abbotsford in the Scottish Borders (together with a number of other extremely odd artefacts, definitely worth a visit). And surely the Islamic world was still technologically superior at the time of the Third Crusade, so why should they not have had better swords?

The idea is so forceful that it prompted a team of German scientists to look for present-day nanotechnology in an old 'Damascus sword' from the Berne Historical Museum in Switzerland. Surprisingly, they found evidence that this steel contained so-called 'carbon nanotubes', a very strong, recently discovered material.[39] They also suggested that a decline in quality during the eighteenth century was due to a change in raw materials, as the ancient Indian mines supplying the ores were becoming depleted, and replacement ores lacked certain impurities essential for the process.

However, a single find in a single sword proves next to nothing. We do not know if nanotubes were indeed a characteristic of all Damascus armouries, nor has it been proved that they were not present in European swords from the same epoch.[40] So the International Society for Research on Damascus Steel will still have many things to discuss, and the mystery, if indeed there is one, of Saracen's Secret remains to be solved.

And what of our spy? Well, Reinhold Angerstein was an iron man from the very beginning, as his family had been for generations before him. He was a very successful and resourceful industrial spy for almost ten years, not only in England and Wales, but also in present-day Belgium, Germany, the Czech Republic,* Austria, Hungary, Italy, France, and Portugal. However, information needs to be acted upon, and it is not clear how much of his meticulously gathered material actually came to be utilized in Sweden.

---

* He did of course visit the Joachimsthal mines in Bohemia (Chapters 1 and 2), although at that time these were known only for their silver production. He noted that the ladies were similarly dressed as in his Swedish hometown of Hedemora.

FIGURE 11 The Marquess of Rockingham's mansion at Wentworth* 1754. Drawing by Engineer Orre in 1760–65 from the (now lost) originals of Reinhold Angerstein's illustrated English travel diary where Angerstein reported that the 'interior is still not quite finished'. Photo by the author from the original transcriptions (1765) in the library of the Swedish Steel Producers Association.

No doubt he would have implemented many things himself, having bought the Vira Iron Works north of Stockholm in 1757—legendary suppliers of swords to the Swedish army. He drafted ambitious plans, but died only three years later, at the age of 41, with most things undone.[41, 42] Ironically, his lasting legacy is to the British people: the diaries from his travels in England and Wales between 1753 and 1755 give many valuable details, not only of industry, but also on many other aspects of life in Britain in the mid-eighteenth century.[43]

Finally, was the association with this notorious Swedish spy an obstacle in the future career of the young nobleman who housed him after his failed expedition to Sheffield? Apparently not. Charles

* Wentworth Woodhouse, as it is known today, is one of the UK's grandest stately homes, boasting the longest country house façade in Europe due to the extensive additions commissioned by Charles Watson-Wentworth.

Watson-Wentworth served twice as British Prime Minister, the first time at the age of 35. He too died in the middle of his active life, during his second stint as PM. He appears to have been very close to his wife, Mary Bright, who acted both as his secretary and political adviser.[44]

5

Biopiracy: The Curse of the Nutmeg

*In this chapter, the chemical side of the worst ever real estate deal is
revealed, we are introduced to spicy molecular siblings, and learn to
draw like chemists.*

Governments and private donors often try to control public
research by handing out very specific grants, expecting closely
related output such as patents, new companies, and inventions in
the specified directions.

Researchers, in general, vehemently oppose such policies, arguing that much better patents, new companies, and inventions will
result if they are left to their own devices, making decisions on
where to use their spatulas, syringes, and microscopes. Grant
applications are therefore sometimes written using an obedient
language adhering to whatever policies and applications are in
vogue at the time, but with a more or less concealed plan B containing the real scientific questions we think should be in focus.

This is by no means a new phenomenon, and one of the most
flagrant misuses of a research grant must have been that of Captain
Henry Hudson in 1609. Issued with a ship, men, and provisions by

the Dutch East India Company (VOC, Vereenigde Oost-Indische Compagnie), the agreed research plan was to explore a route to the Indies by sailing north of Scandinavia and Russia—the so-called north-east passage. He did make an attempt, but somewhere east of Scandinavia's northernmost point, close to North Cape, he had a better idea and turned his ship west. He crossed the Atlantic and, among other things, explored what was to be named the Hudson River. This gave the Dutch Republic a claim to a large island called Manna-hata by the local population, one suspects much to the regret of Hudson's English compatriots.

This urge to go east was partly driven by the enormous profits there were to be made in the spice trade—both on returning home, and on shipping items such as cloves, pepper, and nutmeg within Asia. In a way one can (being a bit chemo-chauvinistic) regard the spice trade as a chemical trade, as a number of very specific molecules make up our sensation of spices compared to the experience of eating rice for example, another important part of the East-Indian trade.

To a first approximation, rice is a mixture of very big molecules such as carbohydrates and proteins, and factors like texture and water content are also important for the overall eating experience. Compare this to cloves for example, their characteristic smell once lingering all over Zanzibar and its archipelago.* If you extract the oil from the dried flower buds of the clove tree it will be almost exclusively, up to 95 per cent, composed of a single substance, a

---

* 'Then, as now, a whiff of cloves and tropical spices came out to greet the traveller from the shore, and on the shore itself a slow, oily sea of marvellous blue washed up on to white coral beaches.' Alan Moorehead, *The White Nile*, Harper & Row, 1960, writing about John Hanning Speke and (Sir) Richard Burton arriving for their 1856 Great Lakes expedition.

molecule with the name *eugenol,* and our sensations when smelling or eating foodstuffs containing cloves is exclusively due to this chemical and a few other related molecules.

This is sometimes confusing, as amply illustrated by an episode from a science broadcast by the Swedish National Radio not so long ago. A team of psychologists wanted to demonstrate the powerful connection between scents and our memories of places or events in our lives (the olfactory memory), the most famous example being when Marcel Proust's alter ego tastes the Madeleine cakes served by his mother and is mentally transported back to the forgotten days of his childhood—the starting point of the novel suite *Remembrance of Things Past.*[45]

But the Swedish psychologists wanted to be a little bit too clever. They gave their experimental subject a chemical sample from a dentist surgery to smell. 'Christmas' was the guinea pig's immediate response, because of the part played by cloves in the making of Swedish gingerbread men around Yuletide. 'Aha!', said the psychologists, 'there you got it wrong, this is a compound called eugenol, used as a mild anaesthetic and antiseptic in dentistry, nothing to do with cloves at all!' And of course it does say eugenol on the bottle, it does not say steam-distilled extract of cloves, so how is one to know? Chemicals, surely, come from factories, not from trees.

Indeed, you probably do not learn things like this in school chemistry anyway, and for all the mild mocking of muggles that goes on in chemistry circles, we should perhaps ask ourselves how much effort we really make in rectifying this situation. I sometimes feel I am teaching at Hogwarts School of Witchcraft and Wizardry, giving away the secrets of an ancient craft few people are even aware of, except for a general feeling that it is 'bad', much like wizardry, and that many of us, like the witches and wizards in Harry Potter, like it that way.

One should add, however, that university student laboratories are, as a rule, more pleasant places than the dungeon where professor Severus Snape teaches his potion classes. On the other hand, there are things that are similar. Students at Hogwarts need to learn runes and other graphical representations of things magical; chemistry students need to learn the graphical language of chemistry, because chemists communicate in pictures and drawings almost as much as in words.

To begin with, we have two languages. A shorthand variety for everyday use in which compounds and materials are given nicknames—or as we say, *trivial names*—such as eugenol or its chemical sibling *isoeugenol*, a component of the most exclusive spice shipped out from the east, nutmeg. These short and handy designations have a distinct drawback, as there is no way of connecting the name to what the molecule looks like or what its formula is, except knowing it by heart. Instead, we can call these two chemicals '2-methoxy-4-(prop-2-enyl)phenol' and '*trans*-2-methoxy-4-(prop-1-en-1-yl)phenol' which will enable most chemists to work out what kinds of molecules we are talking about. Which is fine, if a bit cumbersome, in writing, but of course literally unpronounceable especially when parentheses are needed. This is why we love to draw, and in the process the crafting of the picture becomes a way of thinking too. Figure 12 shows how eugenol and isoeugenol are illustrated.

When I went to engineering school we still had technical drawing classes, making blueprints for reactors complete with exploded sections and all, learning a very efficient and aesthetically highly stylized language of communication. We never had any formal

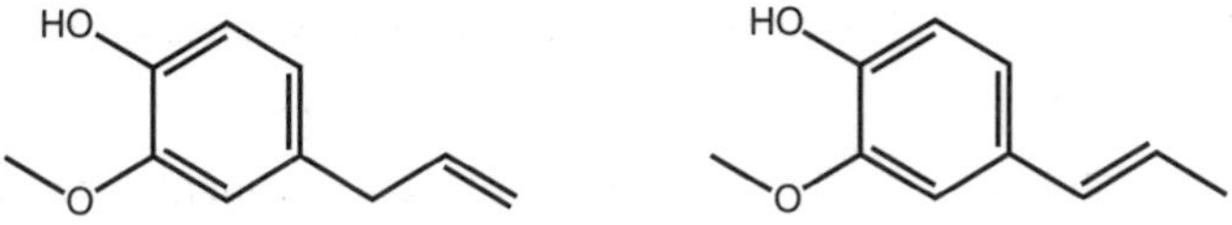

2-methoxy-4-(prop-2-enyl)phenol, better know as *eugenol*   trans-2-methoxy-4-(prop-1-en-1-yl)phenol, better know as *isoeugenol*

FIGURE 12 Different ways of communicating molecular information, in names and in drawing, about the main 'spicy' component in cloves (left) and one of the important components of nutmeg (right), both with the formula $C_{10}H_{12}O_2$.

classes in chemistry drawing, but it was still made clear to us that sloppy drawing was seen as a sign of sloppy thinking.

So, the angles in the hexagons in Figure 12 should be exactly 120°, the second line of the double bonds in the ring placed inside, not outside, and most important of all, there are carbon atoms at all nicks and junction points, but one must never, ever, write out the symbol 'C', and never draw hydrogen atoms bound to a carbon.

Nutmeg contains a mild psychoactive compound related to the eugenols called *myristicin*, which you can see drawn in two different ways in Figure 13. I hope you agree with me that the drawing to the right is very ugly indeed, although it does convey more information to the layman.

It is not myristicin alone that makes nutmeg an attractive spice—in fact, the concentrations are so low that with the small amount of ground nutmeg you generally add to a dish (it goes well with potatoes and spinach, for example) it will have no hallucinatory effect whatsoever, and larger amounts will only make the food inedible.* No, it is the complex mixture of a number of molecules that make nutmeg such a hit in the kitchen.

---

* A Swedish foodies journal published an apple pie recipe a few years ago which carried the misprint of '20 nutmegs' instead of '2 kryddmått' (approximately half a

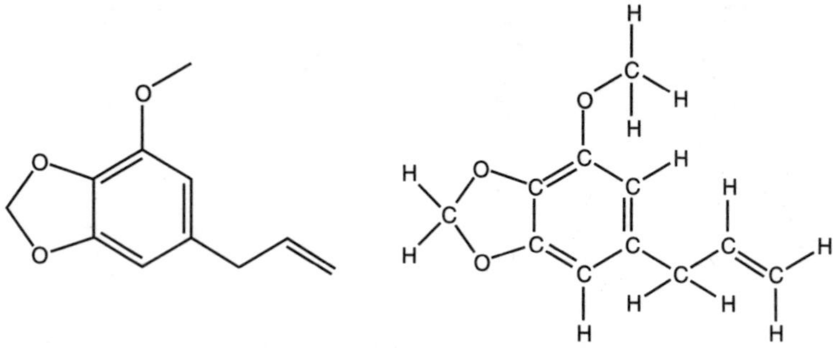

6-allyl-4-methoxy-1,3-benzodioxole better know as *myristicin*

FIGURE 13 Different ways of drawing myristicin with formula $C_{12}H_{12}O_3$. The nice way (left) and one of many ugly ways (right).

If myristicin can be thought of as a cousin of eugenol and iso-eugenol, these two latter molecules clearly must be siblings, having the same formula $C_{10}H_{12}O_2$. Chemists call such molecular siblings isomers.

There are many types of isomer in chemistry, and eugenol and isoeugenol are isomers of the most basic type. They differ by their chemical bonds, as you can see from the different position of the two parallel lines (the double bonds) in the 'tail' on the right-hand side of these two molecules in Figure 12. Just like half-siblings, such isomers may or may not be closely related in ways other than the chemical formula, but in this case they are indeed very close and resemble each other greatly.

What would then be the chemical parallel to real siblings? Molecules with the same formula and all bonds equal, but where the three-dimensional forms of the molecules are different. Isoeugenol

teaspoon). The pie, for those who unwittingly followed the instructions, should have been inedible, but still a few people managed to digest enough to give symptoms of headache and dizziness.

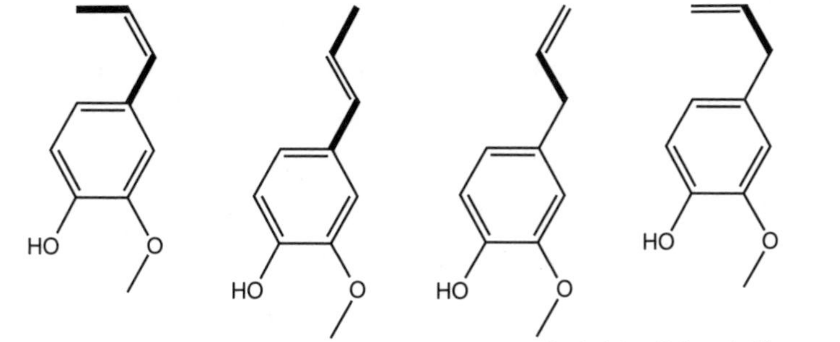

Cis and trans, two isomers, two different molecules

Rotation around a single bond is the only difference between these two drawings that both depict the same molecule

FIGURE 14 Isoeugenol molecular siblings to the left, and two pictures of the same eugenol molecule to the right. Boldface bonds emphasize the difference between cis and trans isomers.

is such a molecule, with the complete name 'trans-2-methoxy-4-(prop-1-en-1-yl)phenol', and it has a brother or perhaps sister formally addressed as 'cis-2-methoxy-4-(prop-1-en-1-yl)phenol'. Less ceremoniously we know them as the trans and cis isomers, where trans has the single bonds on opposite sides of the double bond, while cis has them on the same side, emphasized in boldface above left in Figure 14.

The trans and cis here are exactly the same trans and cis as in the much-debated story about trans-fats, the less healthy fats that you sometimes find in industrially processed cookies and foodstuffs because of the partially hydrogenated fats that they may contain. Not that the trans label is a 'death-eater mark' of any non-naturally occurring molecule. Many natural products contain trans isomers—the essential vitamin A comes to mind—and there are also some natural trans-fats in cow's milk.

As you may imagine, there is a wealth of information in these drawings, well hidden before you have the key, much like medieval paintings that may be pleasing to the eye but where the true meaning is lost without knowledge of the underlying symbolism. The lines that represent the chemical bonds look solid and rigid, like spokes, but in reality molecules are floppy and the atoms are never still; they vibrate and rotate, so in your mind you need to replace the spokes with springs of different stiffness. Take the double bonds: they are double so they are stronger, and you'd think that we could have just made them thicker and shorter, but there is a real mechanical symbolism here. If you have two balls connected by a spoke you can rotate each of the balls independently, but if they are connected by *two* spokes this is no longer possible.

But atoms are connected by electrons, not spokes, and it is not (indeed, it should not be) obvious why this mechanical analogy works. But it does, and that's how we will leave it, because otherwise we would have to spend the rest of the book on quantum chemistry—a fascinating part of the chemical sciences, but not for the faint hearted. Let me just give you the single most important result of quantum chemistry: electrons do not move around atoms like planets around a sun. Some of them appear to do something of that kind, but others move* in very specific, often mutually exclusive, directions, and this is what causes the double bonds to be stiff and forbid rotation. This is the reason isoeugenol comes in two isomers, both naturally occurring, and eugenol in only one.

None of this was known to the two East India companies, the Dutch and the English, that fervently, and often with excessive

---

* OK, so I am not really allowed to say this. In fact, I shouldn't be talking about electrons in this frivolous way at all. I should talk about *electron probability densities* and *wave functions*. This is because electrons, rather like individual animals in a dazzle of zebras, are impossible to distinguish from each other in this way.

violence, competed over control of the spice trade to Europe. One could try to build a case that the fighting over the Banda Islands, the only place where the nutmeg tree was found, could have been avoided had this chemistry been known.

It is very easy to grow the clove tree almost anywhere in the tropics, and if you take the dried cloves and extract the oil you can—in one easy reaction using caustic potash (KOH)—convert eugenol to isoeugenol and voilà, you have transformed the major ingredient of cloves to one of the defining molecules of nutmeg, with no more need for the hard-to-grow nutmeg tree! But it is not that easy. As with many tasty things, the flavour of nutmeg comes from a complex mixture* of compounds—cloves are rather an exception, relying almost exclusively on one molecule for effect.

So, simple chemistry would not have saved lives here. The brutal fighting over the Banda Islands, especially one of the smallest, Run, was instead to end with the Treaty of Breda in 1667 after the second Anglo–Dutch war. As a small item in these negotiations, Run—the only 'English' of the Banda Islands, conquered by the Dutch after a four-year siege—was to be kept by the Netherlands, and in return the English got to keep Manhattan Island on the Hudson River, which they had recently occupied.[46]

In the short term this was very profitable for the Dutch, who now had a world monopoly on nutmeg. This was, however, constantly threatened by what today we call biopiracy. Of course, the Dutch were unscrupulously profiting from what were rightfully the Banda inhabitants' resources, but the 15,000 islanders could

* To make it even more complicated, this mixture will change depending on the variety and the place where the spices have been grown.

now perhaps be living nicely on the around 80 million US dollars the nutmeg export is worth every year, had the British not managed to plant nutmeg trees in other parts of the tropics in 1817.[47]

Today, with the United Nations Convention on Biological Diversity in force, moving biological material this way would be a blatant violation of the agreement, signed by 150 governments at the 1992 Rio Earth Summit.[48] One of the main issues is to protect both biological resources and know-how about them, especially in poorer countries, from being exploited by foreign interests.

Another reason for protecting the planet's biodiversity is that it is one of our most important molecular resources. You might think that the chemical industry could easily whisk together a process for making any kind of profitable molecule, but this is not true all the time. Synthesizing molecules like eugenol, isoeugenol, and myristicin from 'scratch' is not always viable—natural sources are often used. This will be especially true after we have run out of oil—that wonderful source of chemical starting materials.

There may in fact be many valuable molecules not yet discovered, in plants that are becoming extinct this very minute—cures for HIV, TB, new antibiotics, or the perfect precursor for making a catalyst that converts the Sun's rays into useful energy.

The Dutch East India Company had lost its importance by the end of the eighteenth century, and was dissolved in 1800. Shortly thereafter the Banda Islands lost their economic importance for the Dutch. The property prices on Manhattan continued to rise, however, and are now among the highest in the world. One could argue that even if the American Indians got a lousy deal over this island, the Dutch did even worse, and of course the British played their cards in North America so badly that all was lost anyway.

A lasting effect to this day, however, is that the Netherlands is the largest import market for nutmeg. A vivid account of the history of this nut, that is botanically speaking not a nut at all, can be found in Giles Milton's book *Nathaniel's Nutmeg*.[46]

6

Death at Number 29

In this chapter we meet individuals high on 'T', and learn about the removal of electrons from copper and people from high places.

The five others went first, one by one, and contemporary sources noted how humane the spectacle was, as the participants did not need to see each other. Thousands of Stockholmers had turned out to watch, on this cold day of 30 January 1744, as the last of the six, Gustaf Schedin, accountant at the Insjö copper works, mounted the scaffold. As the culmination of the show, he would be both beheaded and then cut to pieces.[49]

The summer before, Schedin had led the fourth Dalecarlian Rebellion: the last march of the free miners and farmers of Dalarna— the mine-rich county 100 miles north-west of Stockholm—to the Swedish capital, in a movement expressing raging discontent with the king, Fredrik I, and his disastrous war with Russia. This sort of thing had been successful before: the fiercely independent-minded people of Dalarna traditionally wielded a certain power, rich as they were in natural resources—the jewel in the crown being the famous Great Copper Mountain mine in Falun. Once it was the largest of its kind in the world, and yielded something like 70 per cent of the world's copper production.[50]

The Falun mine, like many others, was once managed as a cooperative operation, and worked by free miners called mountain-men (bergsmän) with special privileges and laws of their own. But their time was at an end. In 1743 the uprising ended in a bloodbath in Stockholm, and now the six leaders were to be executed. The copper mine was also losing its privileged position. It had given the Swedish kings and queens economic strength for numerous more-or-less successful military adventures down in continental Europe, but was now in decline, and so was the military power of Sweden.

This traditionally male activity—becoming angry and getting the lads together to sort things and people out—is chemically related to high levels of the large organic molecule testosterone. For a inorganic chemist inclined to find a good story, it would have been great to now present a direct link between copper and the way we make this molecule in our bodies, starting from cholesterol, claiming that this made the men from Dalarna more inclined to hasty revolutionary actions. Well, forget that idea, because there is no trace of copper in this chain of events. Iron and zinc are essential to some of the enzymes—the proteins that catalyse these reactions in the body—but not copper, as far as we know, although copper performs the role of reaction centre for many other enzymes. Not much to build a case on then, so on we go to discuss more examples of political turmoil linked to copper mining.

I am too young to remember 1962, but have some faint recollections of 1973—the famous last picture of President Allende in the portal of La Moneda palace in Santiago, helmet on his head, and the young man in a suit and tie with an automatic rifle in his hands. In 1962 there was the crash at Ndola in present-day Zambia—photos of the wrecked DC-6 and Northern Rhodesian officers in shorts. Less remembered in the West is democratically elected Congolese Prime Minister Patrice Lumumba, killed in custody under the Mubuto regime in 1961, but with the silent approval of the Belgian government (an act they officially apologized for in 2002).[51]

Power and wealth—but also war, murder, and all sorts of violent conflicts—are intimately bound to the two heavier metals in the triplet column called the coinage metals, silver (Ag) and gold (Au), and gold has even been called the most poisonous of all elements, particularly to the mind. We have all seen images of pirates fighting over a chest full of gold doubloons, but there never seems to be any swashbuckling over trunks of copper pennies. Yet copper was arguably at the heart of the conflicts in Chile and the Congo that claimed the lives of President Salvador Allende and UN Secretary General and Swedish diplomat Dag Hammarskjöld.

Both men constituted real or perceived threats to powerful copper mining companies. Hammarskjöld was trying to resolve the conflict over the breakaway republic of Katanga in the southeastern corner of the newly independent republic of Congo, and Allende, as the first elected socialist president, had nationalized Chilean copper mines a few years earlier. While Allende almost certainly committed suicide during the coup d'état and the attacks on the presidential palace,[52] the aeroplane crash that ended the life of Hammarskjöld and 15 other passengers and crew is still disputed, and new facts have been brought forward, notably after the fall of the apartheid regime in South Africa.[53]

A mine is a hole in the ground with a chemical factory on top. This is especially true for copper mining and production, as very often not one, but two, important products—one solid and one liquid—are trucked out from these plants.

Just as in our enzymes, copper is not normally found as bright shining copper metal in nature. Instead, having 11 electrons out of a total of 29 in its outer shell, it has usually lost two electrons and become copper(II) or Cu^{2+}, or lost one electron and become copper(I) or Cu^+. This is convenient for us, as these ions are often water soluble and easier to digest than solid chunks of copper plate.

In our body, the copper ions are looked after mostly by nitrogen atoms from the proteins (one amino acid, called *histidine*, is especially good at this), but also by the sulphur-containing amino acids *cysteine* and *cytosine*. To reveal a little secret, copper has a perpetual love affair with sulphur—a fact our enzymes use to make copper do some very important reactions in our body. This love affair is manifested not only in its love for amino acids with sulphur, but also in the greenish-blue patina acquired by copper statues (notably there are none of Fredrik I—best forgotten seems to be the common verdict) and other copper objects exposed to the elements and to sulphur pollutants.

The copper–sulphur connection is also found underground, as many copper minerals have sulphur in them: for example, chalcopyrite ($CuFeS_2$). As copper works grew in scale, separating the loved ones became an environmental hazard. This was because the process uses oxygen from the air that, together with the copper 2+ ions, take two electrons from each sulphur atom. This reaction transforms copper(II) to copper metal, and the sulphide ions (S^{2-}) to sulphur dioxide or sulphur trioxide molecules (SO_2 and

SO_3), where the sulphur atom is now well hidden inside a shell of oxygen atoms and can no longer feel any attraction for the copper. Naturally SO_2 and SO_3, both being gases, were just let out of the chimney—problem solved!

$$3Cu_2S + 3O_2 \rightarrow 6Cu + 3SO_2$$

But, as has happened many times before, one man's waste is another's gain. Instead of releasing the SO_2 and SO_3 into the air, where they would combine with water vapour and form sulphurous acid (H_2SO_3) and sulphuric acid (H_2SO_4) and cause acid rain, these gases can be collected and turned into 100 per cent SO_3 through oxidation with O_2 from the air, with the help of clever new catalysts. Water is then added, and we have produced sulphuric acid again, but this time we can sell it. And there is quite a market—year after year sulphuric acid heads the top-ten lists of chemical products worldwide.

Sweden today accounts for only 1 per cent of the world's copper production, and the Falun mine was closed for production in 1992. It is now a UNESCO World Heritage site and can be visited both above and underground. The original company of mountain-men in time became the Stora Kopparbergs Bergslags Aktiebolag (Great Copper Mountain Miners Guild Inc), arguably the oldest incorporated enterprise in the world. They merged with Finnish Enso in 1998 and are nowadays into much softer things, as Stora Enso is one of the largest pulp and paper companies in the world.

7

Blue Blooded Stones and the Prisoner in the Crystal Cage

In this chapter we learn about the ancient art and science of crystallography, a discipline not only relevant to your jeweller, but found in your kitchen and practiced by materials scientists, drug developers, and biochemists alike.

You have no doubt heard about blood diamonds, and know that they are not rare red versions of the gemstone, but illicitly mined diamonds used to finance and prolong armed conflicts in some African countries. But have you heard of blue blooded stones?

An elaborate marking system known as the Kimberley Process Certification Scheme is currently used, although some claim inefficiently, to sort good diamonds (for example, from Botswana) from blood diamonds that should not be allowed into the market. No such scheme is needed for the blue stones named *lapis lazuli*, as there is only one mine in the world that produces high-quality stones—the Sar-e Sang mine in the Kokcha valley in the

Badakhshan province in north-eastern Afghanistan—so there is never any doubt about where they come from.[54]

The mine is in such a remote area that even prolific travellers like Marco Polo and Sir Richard Burton never made it there, although Polo refers to them in his travels when crossing the river Oxus (also known as the Amu Darya) of which the Kokcha is a tributary: 'a mountain in that region where the finest azure in the world is found.'[55] A Scottish explorer, John Wood, visited in 1837, but if his book *Journey to the Source of the River Oxus* is to be believed, it wasn't exactly a Sunday School excursion either: 'If you wish not to go to destruction, avoid the narrow valley of Koran [Kokcha],' he summarized.[56]

One who finally made it there was the British journalist Victoria Finlay, author of the wonderful *Colour: Travels Through the Paintbox*, and, although reaching the mine in the beginning of the 2000s, this was still quite an achievement.[54]

Why would anyone endure various kinds of hardships just to see a mine where you can whack out blue stones from the interior of a mountain? Perhaps because these rare stones have achieved tremendous value over the ages, being the hallmark of kings and aristocracy, or because the trade in them covered such distances even in ancient times, or maybe because this mine is possibly the oldest in the world that is still being worked, having been in business for 5,000–6,000 years. Finlay's reasons were not so much the stones themselves as the pigment they produce when ground to a fine powder, forming the base of the best and most expensive blue colour—natural *ultramarine* (the word ultramarine meaning from the other side of the sea). We now have synthetic versions of this pigment, but they are still not quite of the same quality, and as we shall see, there is a very good chemical reason for this.

Afghan politics is as convoluted as a detective novel by Raymond Chandler, at least when viewed from a distance, so whether the next person in this story is still seen in such a positive light in the years to come remains to be seen. Ahmad Shah Massoud, the 'Lion of Panjshir', is an official national hero in Afghanistan, and on the day of his death, 9 September, many Afghans make the pilgrimage to pay tribute to him at his grave in the small village of Jangalak.[57]

While an engineering student in Kabul in the 1970s, he became politically active and had to flee to Pakistan, returning a few years later to become a legendary mujahedin commander during the Soviet occupation. Massoud was so successful that the Soviet army, superior in arms and manpower, never had full control over the small province of Panjshir, just south of Badakhshan, where the lapis lazuli mine is situated. He probably exploited the mine, but it became a true lifeline after the Afghan civil war that saw the Taliban triumphantly enter Kabul and Massoud leave his post as minister of defence and once again take to the mountains.[58]

We hear a lot about the opium trade financing the Afghan civil war, but it seems the Northern Alliance led by Massoud got most of its money from controlling the international trade in lapis lazuli and other precious stones, mainly emeralds, ironically used to buy weapons from their former Russian enemies.[59] And like the Russians, the Taliban never got complete control over these provinces. Massoud was a devoted Muslim himself and something of a scholar if not a spiritual leader, and rejected the Taliban's fundamentalist views in favour of a more modern version of Islam.

So what is it that makes these blue stones so special? First, as with all gemstones, it's about their childhood. Growing up and reaching adulthood too fast is not good, either for humans or for

FIGURE 15 Ahmad Shah Massoud, the 'Lion of Panjshir'. © Reza/Webistan/ Corbis.

gemstones. These crystals, as chemists would call most gemstones, grow up from tiny assemblies of molecules and ions called seeds. Nobody really knows how small these are—perhaps a few molecules or a few hundred, swimming around in some kind of liquid or melt. Once in a while a new molecule from the liquid will hit the surface of such a seed and get stuck, making the seed grow bigger.

Now, the thing about crystals is order. Take a sugar crystal from your kitchen and look at it. It looks like a shiny little gemstone, it is clear and transparent and it has nice, well-defined edges and

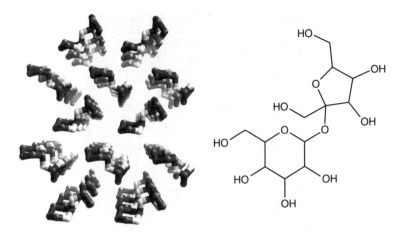

FIGURE 16 Left: sucrose molecules on parade in a crystal of sugar. For simplicity only the two rings, one with five and the other with six atoms are shown. The complete formula is $(C_6H_{11}O_5)O(C_6H_{11}O_5)$. Right: simplified line drawing of the sucrose molecule.

faces. It has these characteristics because inside the crystal you will find the sucrose molecules all lined up in the same way—up, down, left, right, front, and back—like soldiers on parade, vines in a vineyard, or oranges neatly stacked in the fruit shop. It is just that there will be many more of them than you would ever care to imagine.

If soldiers arrive for a parade at a leisurely tempo one by one, it is easy for them to find their place in the grand scheme of things. But if they all arrive on the showground at the same time, running in from all directions because they were late for breakfast (when I was in the army I had a friend who excelled in this type of behaviour), it is hard for them to form the militarily precise array that is expected of them. Possibly they can keep track of their closest neighbours, but overall there will be a mess until a sergeant arrives to shout them into order.

Molecules don't have sergeants, they are all equals, so once this process has gone wrong in the molecular world there is usually no turning back. Take your nice-looking sugar crystals from the pantry, dissolve them in hot water, and let the water rapidly evaporate again. This makes the molecules rush back together to form a solid, and you will get back your sucrose molecules in a solid state, but gone are the beautiful crystals! You may get anything from a gooey mass to malformed crystals, depending on the exact conditions.* Come to think of it, you don't find nice pieces of marble stone in the bottom of your tea kettle either, although in principle it is the same calcium carbonate in the deposit that forms around the heating coils (depending on where you live and the hardness of your water). Table salt is easier to get back into crystalline form, so obviously this depends on the exact molecules you have, but in general *crystal growth* is a tricky field with few rules and only weak guidelines. Consequently, some of my chemist colleagues jokingly refer to it as a 'black art', and yet others invoke the actions of the crystal fairy in their test tubes.

Why do we care so much about crystals? Because they help us see the molecules we work with, something that has helped the progress of chemistry, medicine, and materials science tremendously since we found out how to make such images about 100 years ago. Molecules are generally too small for even the most weird or sophisticated type of modern microscopy,† but when they

---

* There *are* good ways to grow sugar crystals from a water solution. With a bit of patience, as you need days, these are good experiments to do with young children, but then the reward is that you can grow really big ones!

† Good images have recently been obtained for flat molecules using Atomic Force Microscopy (AFM), but most molecules, like sucrose, have a distinct non-flat shape and cannot be investigated this way. J. Svensson, 'A really close look at molecules', *Kemivärlden Biotech med Kemisk Tidskrift*, March, 2013, 39,–41.

stand to attention inside a crystal the very order of things turns out to be a great help: atom–atom distances, inside and in-between molecules, will be repeated millions of times. Your average sugar crystal will contain around 10^{18} molecules, 1 quintillion, or the same number as the solution to the famous grain and chessboard problem.*

If we aim a narrow X-ray beam on the crystal, the photons in the ray will bounce off the atoms—or rather their electrons—that they hit. Some at the surface, some a layer down, some a layer further down, and so on. If we place a photographic plate (or some other detector device) behind the crystal, in the centre we will detect a big spot from the beam that passes right through the crystal, but also weaker spots surrounding it. These off-centre spots come from the photons that have hit an atom on the way. Because these waves of light have travelled different distances, the peaks and troughs of the wave are now in different places from the rays that were reflected from the top layer and the layer below that had a slightly longer distance to cover.

For normal light this is such a small discrepancy that it does not matter, but as X-rays have wavelengths that are close to the atom–atom distances both within and between molecules, the difference in path may be as much as half a wavelength. When such combinations of rays emerge from the crystals and recombine to form the reflected beam, a peak in the wave from one reflection is now combined with a trough in the wave from another reflection—the two waves cancel each other out and

---

* The fable where the inventor of chess is asked by the pleased ruler to name his reward and replies that he wants one grain of rice on the first square of the chessboard, two on the second, then again doubled on each square until the sixty-fourth.

there will be no spot on the film. This is known as destructive interference.

On the other hand, if the difference in distance travelled is one, two, three, or any whole number of wavelengths, the photons will combine peaks with peaks and will be 'in phase'. This is constructive interference, but crystallographers will simply say that in those cases you see a *diffraction peak*. By measuring the intensity and position of these peaks (called a diffraction pattern) on a photographic film, the Braggs (father and son) were able to determine the positions of the sodium and chloride ions in table salt crystals in 1913, and thus became the first two people ever to 'see' an atom.[60] Nowadays we often use CCD detectors, like those used in digital cameras, to record these spots.

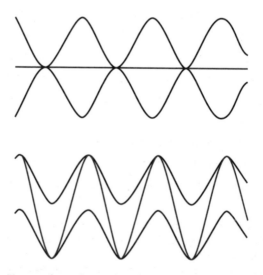

FIGURE 17 Top: peaks and troughs of waves that combine exactly half a wavelength out of phase cancel (grey line) and give a zero signal. Bottom: in-phase combinations are reinforced and give a strong signal (grey wave).

Despite this talk of peaks and troughs, we have wandered far away from the Afghan mountains. I have done this so that you can appreciate the true beauty of lapis lazuli (or rather the mineral lazurite), the magnificent blue component of the lapis lazuli rocks, because this material is infinitely more delicate and complicated than that king of gems, the diamond. Already, before World War I, the Braggs could work out the positions of the carbon atoms in diamond with photographic film, paper, and pencil, while crystallographers are still struggling with the structure of lazurite.

Lazurite is complicated because its chemical formula tends to vary. Compared to diamond, which is simply 'C', we usually write lazurite as '$Na_6Ca_2(Al_6Si_6O_{24})[(SO_4), S, Cl, (OH)]_2$' where the exact amounts of the last four components will vary from rock to rock.* Why does this chemical behave in such a funny way, not abiding by the rules we have learnt in school about strict proportions of the elements? The reason is the three-dimensional structure of the $Al_6Si_6O_{24}$ part. This is fairly straightforward to work out from the X-ray diffraction pattern, and what you see then is an infinite interconnected structure composed of AlO_4 and SiO_4 tetrahedrons, where every oxygen is bridging an Al/Si pair of metal ions.[61, 62]

So what, you might say—if you paid attention in chemistry class, diamond is also based on tetrahedrons. But in this case the distances between the centres of the four-connected network are much longer, and create huge cage-like pockets of empty space

---

* As an example explaining crystallography, this compound is actually not the best, as it is a bit more complicated than your average crystal. You can think of the variable composition as soldiers on parade wearing a few different caps, creating another kind of order in addition to the shoulder-to-shoulder men in neat rows and columns. The technical term is 'incommensurate modulated' and my crystallography friends happily tell me that this is not a problem to understand, you just move into the fourth or fifth dimension of space.

inside the crystal. In these we find the relatively boring small and uncoloured sodium and calcium ions occupying smaller pockets, and the likewise transparent but bigger sulphate, chloride, and hydroxide ions locked into the larger cages. What about the colour then, does that come from the silicate-aluminate framework? No, such a framework forms the basis of a number of minerals. The key is perhaps to be found in another component making up the lapis lazuli: the presence of small crystals of iron pyrite or *fools' gold* speckled around the rock. This fools' gold not only adds beauty to the stones, but may also have been important for the formation of the real prisoners of these crystal cages—the negatively charged S_3^- ion[63]—as the pyrite has the formula FeS_2, and contains S–S units that may once have been the starting point for the making of the S–S–S⁻ ions.

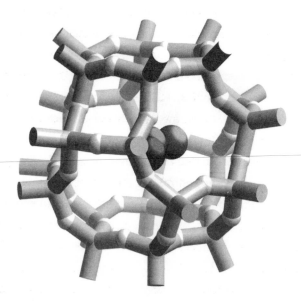

FIGURE 18 The lazurite crystal cage of Al–O and Si–O bonds and the trapped S_3^- radical ion inside, that gives the blue colour to lapis lazuli gems, shown by three darker spheres.

This unusual species is present only in very small amounts, and needs to be kept in the cage as it is a radical. (Chemistry is not really so conservative that all radicals need to be kept under lock and key, it is just that we call a molecule that contains unpaired electrons a radical, and such fellows are normally very reactive.) You have probably heard of the 'free radicals' doing all sorts of damage to your body. These are the OH radicals; the S_3^- ion is another species of the same kind. It gives the blue colour to the Lapis lazuli, but if you let it out of the cage it will immediately be destroyed and the colours vanish. Part of the secret of making a good paint out of lapis lazuli is probably to make sure that this does not happen during the process.

The cages and their prisoners are also the reason why it is difficult to reproduce these materials in the laboratory or on an industrial scale. The S_3^- ions are too big to get out of their cages if these are intact, which also means it is impossible to just put things inside the cages. It has to be an integrated process, with the unstable S_3^- ions forming at more-or-less the same time as the cages are built around them. How they achieve the conditions for doing this, to give the same quality gems as found in nature, is yet to be worked out by chemists.

If it is done, however, that may harm future rebels taking to the Badakhshan mountains and relying on this trade for income. For Massoud, it ended badly anyway. In what is considered to have been a prelude to the 9/11 attacks, he was assassinated by two Al-Qaida suicide bombers on 9 September 2001 at Khwaja Bahauddin, in the Takhar Province in the north-eastern part of the country. Since 2005, 9 September is known as 'Massoud Day', and is a national holiday in Afghanistan.[57]

8

Diamonds are Forever and Zirconium is for Submarines

In this chapter we learn how the relative positions of the elements in the Periodic Table can help us predict not only their properties, but also where to find them in nature. We also talk about nuclear reactions, fake diamonds, and the major media drama of 1952.

The appearance of a diamond engagement ring in the long and convoluted love story between Botswana's First Lady Detective, Mma Ramotswe, and the owner and brilliant mechanic of Tlokweng Road Speedy Motors, Mr J. L. B. Matekoni, seems to signal an end to this particular sub-plot, stretching over several volumes of Alexander McCall Smith's bestselling and original series of crime novels (that we met in Chapter 1). However, a slight problem involving cubic zirconia is discovered, and the story lingers on until the next book in the series.[64]

Similar names for elements and their compounds are a nuisance in chemistry,* but often arise historically, and zirconium is just one

---

* I do not want to give you the impression that chemists are sloppy in their name-giving practices—these are more like nicknames, or trade names. The

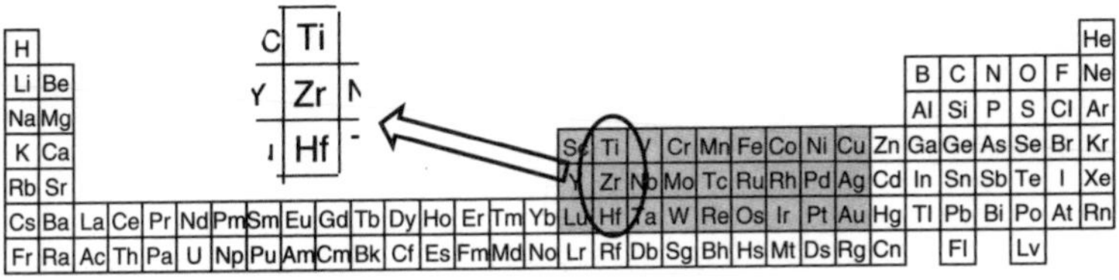

FIGURE 19 The Periodic Table, with the transition metals in grey and the titanium group enlarged.

such example. Apart from the pure metal we have *zircon* and *zirconia*, all three of which have important applications. Zircon is zirconium silicate, with the formula $ZrSiO_4$, and cubic zirconia is a special form of zirconium dioxide, ZrO_2. The latter, as you may have guessed, is an excellent diamond substitute in, among other applications, engagement rings.

We are not going to dwell on the details of the element zirconium, but you should know that within the Periodic Table it is located in the large middle chunk called the transition metals. You have probably heard of its cousin titanium, immediately above it, and a sibling, hafnium, straight down the ladder.

Why do I call them siblings? Because in the Periodic Table elements in the same column tend to have similar chemical properties. In particular, in the family of transition metals in the central section containing 27 elements—each with a number of properties in common—the two lower elements in each column tend to be the most similar.

The similar chemical properties of zirconium and titanium means that we can usually find zirconium where we mine the

International Union of Pure and Applied Chemistry (IUPAC) is continuously developing terms, words, and grammar for the unambiguous naming of chemical compounds. This tremendously important work enables trade, control, legislation, and customs to work properly around the world.

much more plentiful titanium, and also that once we have separated the titanium from zirconium there will be a small quantity of hafnium trailing along—an impurity that is much harder to get rid of.

The sleek jeweller in Gaborone will not care if his fake diamonds contain trace levels of HfO_2 mixed with the ZrO_2. This will not affect the lustre, hardness, or transparency to the untrained eye, but for the engineers constructing the first power-generating nuclear reactors in the US it was a very different matter.

Extensive material trials after World War II showed that an alloy with zirconium metal as the main component was the best candidate for enclosing the uranium oxide in the fuel rods for use in nuclear reactors. The snag for the engineers who had to make this work in a power plant, however, was that the metal needed to be completely free of hafnium. The reason is that the similarities between zirconium and hafnium end at the nucleus, where chemistry stops and physics begins. These two elements (or to be more precise, the naturally occurring isotopes of these elements) react very differently if hit by a neutron, or, as a physicist would explain: their *neutron cross sections* are very different.

A steady flow of neutrons is needed in the nuclear reactor to drive the uranium fission chain reaction at a low and suitable speed. Too many and the reaction will run wild, too few and it will just stop. What happens in a conventional nuclear reactor is that uranium-235, or, as a chemist would write ^{235}U, is hit by a neutron with mass 1. Then many things can happen, one of the more important is that the uranium nucleus splits and forms two new atoms, ^{92}Kr and ^{141}Ba. As you may notice, these two new mass numbers do not add up to $235 + 1 = 236$; there are three units missing. That is because three new neutrons are sent out during the reaction, each one able to split another uranium nucleus and produce three new neutrons. This is the basis of the famous chain reaction behind the atom bomb. The reaction can go berserk if the number of neutrons on the loose in the reactor is not carefully controlled.

During the atomic energy project, neutron capture ability was just one of the many properties that materials needed to be analysed for, to a level and detail unprecedented in any engineering project before. It was found that all the natural isotopes of zirconium have a very small thermal neutron capture cross section, meaning that a neutron that hits a zirconium atom will just bounce off and continue to another atom. The hafnium isotope ^{178}Hf on the other hand, that makes up every second naturally occurring hafnium atom, has a high liking for these elemental particles. ^{178}Hf will easily absorb a neutron to give ^{179}Hf, which is also stable and naturally occurring. Any hafnium in the cover of the fuel rods will quickly absorb neutrons and stop the chain reaction and the reactor will shut down. The upside of this is that hafnium can be used in the *control rods* that can be lowered in-between the fuel rods in the reactor, efficiently stopping the fission process by guzzling up all the neutrons that feed it.

We started this chapter with a wedding ring in Gaborone, close to the Kalahari Desert in landlocked Botswana, but we will close it with a Christmas tale from the North Atlantic. We start with a coincidence: hafnium is named after the Danish capital Copenhagen (Køpenhavn in Danish), and just before Christmas 1951 the legendary (we will soon see why) Danish merchant captain Curt Carlsen left Hamburg with the US flagged freight steamer *Flying Enterprise* bound for New York. Not only was the captain Danish, the ship was owned by the New York-based Dane Henrik

Isbrandsen, an unconventional shipping entrepreneur related to the famous Maersk family whose ships are seen in all major ports even today.

Anyone old enough to follow the news in the early days of 1952 apparently remembers the events that unfolded just after Christmas 1951. One of the most terrible storms of the 1950s had hit the North Atlantic and north-west Europe, and a freak wave hit the *Flying Enterprise*, which was steaming out from the English Channel, and caused a crack in the middle of the hull. The crew made provisional but seemingly solid repairs, and the captain ordered the ship to steam on as best it could, though some of the crew had mixed feelings about this decision and would have preferred Carlsen to head for shelter. Sure enough, a second wave came down on the ship a few hours later and rearranged the Volkswagen cars in hold three, or possibly the pig iron in hold two, giving the ship a 60° tilt that it never recovered from.[65]

Our interest is in another of the holds, where an undeclared cargo of zirconium metal resided, but what first caught the public's attention was the rescue operation. The second wave had also killed the engines and disabled the steering, and Carlsen now had no other choice than to issue the order to abandon ship. This was easier said than done, as the lifeboats were inoperative because of the tilt. Luckily three other ships had answered the mayday call, among these the USS *General A. W. Greely*, so Carlsen made the crew and passengers jump into the water and swim for the approaching rescue boats.

This was perhaps lucky at the time, but in retrospect Carlsen may have wished that the US Navy had been busy elsewhere. The continuous stand-by of a succession of US Naval ships close to his own half-capsized vessel fuelled endless speculations that this old and run-down ship was carrying a very secret and important cargo for the US government—speculations that were to follow Carlsen until his death.

Circumstantial evidence abounds. Why did the US Navy not rescue other ships damaged by the storm? Why did the British Royal Navy also pay close attention? Why was part of the cargo rescued in a secretive operation in the spring of 1953? Why did Carlsen not turn around after the first freak wave? And above all, why did he stay on his sinking ship for almost two weeks until a few short moments before it went down in the English Channel, some 70 kilometres outside Falmouth in Cornwall?

There is a certain unflattering streak in our human nature in that, much as we admire our heroes, nothing gives us more satisfaction than to bring them down—their very human weaknesses making us feel better. Thus, it could not have been classic seamanship, nor the enormity of a captain losing his ship, nor the fighting spirit of someone not ready to give up to the last; it had to be something else that kept Carlsen going. And this 'something else' would have to have been a captain under strict orders from the highest US authorities not to abandon ship, as otherwise the secret cargo could have fallen into the wrong hands. The more cynically minded speculate about a hefty but secret reward awaiting the captain in New York.

Why all the hush-hush, anxiety, and cover-up? Because the zirconium was intended for the reactors in the world's first nuclear submarine, the USS Nautilus. Two decades later, Carlsen sneered at 'stupid journalists' who suggested that he had carried material for nuclear weapons, and said that because of his failure in delivering the cargo, the launching of the Nautilus was delayed for six months.[66]

But how secret was this really? Certainly the nuclear properties of zirconium were not something the US Government was prepared to share with everybody and his uncle, but as early as March 1951 the US Atomic Energy Commission issued a press release announcing the intention to obtain zirconium and hafnium from commercial sources and asking for expressions of interest. In November the same year 35 companies were invited, and in January 1952 six companies made bids for the contract.[67]

This was a deliberate strategy by another legendary sailor, Admiral Hyman Rickover, at the time both the commanding officer of the Naval Reactor Program and a US Atomic Energy Commission official. Having taken the decision to use zirconium in the fuel rods as early as 1947, he did not want the naval reactor programme to be dependent on one supplier only, especially not a government agency, even though the Oak Ridge National Laboratory (of atom bomb fame) and the National Bureau of Mines had been successful zirconium producers in the initial phases.[68]

So, if Carlsen shipped a 'secret' cargo of zirconium it was not necessarily on the orders of the US Navy, but perhaps on behalf of one of the companies involved in the bidding for the government contract.

Where would this zirconium have come from then? It is possible that the Philips Company in Eindhoven had a relatively large stock of pure zirconium, as in 1928 they had obtained the first patent for the separation of hafnium from zirconium and were producing the metal in pure form at least until 1950,[69] mostly for use in photography flashlights.

Carlsen himself claims his zirconium came from the Nazi Germany atom bomb and atom energy project, the *Uranverein*.[66] However, you don't need zirconium to make a crude uranium-based nuclear bomb, only for a power-generating reactor, and the Germans did not make significant advances in either direction. The Nazi regime may just have stolen the stock from Philips anyway, so there is no saying that Carlsen is definitely wrong on this point.

What seems to be without doubt, however, is that Admiral Rickover got the *Nautilus* ready on time, meeting the official timetable in advance, but narrowly missing his own more ambitious target because of a problem with the steam pipes in the non-nuclear part of the power plant. The whole zirconium project was, however, a calculated risk. Not because of a supply problem—the metal is easier to find than many common metals such as tungsten, chromium, zinc, and copper—but because of the engineering difficulties in producing the pure metal. One of Rickover's closest civilian co-workers, Ted Rockwell, told me that it 'was a hard and furious race' that 'could easily have been unsuccessful, right until the last moment'.[70]

But it worked, and the rest, as they say, is history. Rickover* also oversaw the first dedicated peaceful nuclear power plant in Shippingport, Pennsylvania, which was hooked up to the electrical grid a few years after the launching of the Nautilus. Moreover, the standards he set for the 'nuclear navy' has enabled the US Navy to work its power-generating reactors without any accidents so far.[71]

Zirconium is not without problems though. It is a strong and very unreactive metal in normal circumstances, being as stable against corrosion as gold, but if the cooling of the fuel fails,

---

* Rickover was known as 'the Father of the Nuclear Navy', but was by all accounts a controversial figure. A one time US Chief of Naval Operations has been quoted saying: 'The Navy has three enemies, the Air Force, the Soviet Union, and Hyman Rickover'.[67]

and the control rods cannot be inserted to stop the chain reaction, the extreme heat will make the zirconium more like sodium. And now we will have the typical school demonstration we encountered briefly in Chapter 2 when discussing calcium. A shiny piece of sodium metal dropped into a jar of water, whirling around on top of the liquid surface, burning and sometimes producing a bang or a more discrete popping sound.

Reaction 1 : $2H_2O(liquid) + 2Na(solid)$
$$\rightarrow 2Na^+ + H_2(gas) + 2OH^- + heat$$

Reaction 2 : $2H_2(gas) + O_2(gas)$ $2H_2O(liquid) + a lot of heat$

It is not the metal we see burning, but the hydrogen gas generated in the reaction between the metal and water (Reaction 1). Sometimes small explosions occur when the concentration of unreacted hydrogen gas builds up, mixes with oxygen from air, and then reacts all at once (Reaction 2). This is known as a hydrogen explosion, and is just what happened inside the Fukushima (Honshu, Japan) reactor in 2011, and probably in the Three Mile Island accident in 1979 (Pennsylvania, US), only in these cases the metal was not sodium but zirconium.

Finally, if there are any prospective buyers of diamond rings among the readers, they may be interested in how to best avoid J. L. B. Matekoni's problems. Besides the obvious test of lowering the possibly fake stones in-between the fuel rods in a nuclear reactor and seeing if the power output goes down because of the hafnium content, there is actually not very much you can do if the stone is small and set in a ring.

Cubic zirconia and diamond differ slightly in hardness and refractive index, but both scratch tests and see-through tests may be difficult to perform on a diamond ring. The density is also different—carbon atoms obviously 'weighing' less than zirconium atoms—but this is almost impossible to test on a ring.

If you are in the market for a substantial diamond, it is very easy though. Simply hold the stone against a sensible part of your skin, the upper lip for example: a real diamond conducts heat very well; it will quickly drain heat from that part of your skin and so will feel cold, just like a piece of metal, whereas zirconia is an insulator and you will not feel any difference.

9

Graphite Valley: IT in the Eighteenth-Century Lake District

In this chapter we will discover that the pure elements can come in different disguises, and that it isn't only diamonds that are worth smuggling. We will also learn more about chemical bonds and how to make electrons jump.

Lake Windermere in the north-west of England perhaps makes you think of poets, or of adolescent adventures less concerned with wizards and vampires and more with Swallows and Amazons if you have grown up with English children's books. Anyhow, people who lived by their pencil. Or should that perhaps be the pen? We don't see the serious author in her study hard at work with a pencil. Pencils are generally considered to be mostly for children doing their homework, or others who frequently need to erase their mistakes.

There has never been a lack of ink, traditionally a mixture of iron salts, water, and tannins—the bitter tasting compounds in tea*

---

* To demonstrate this, first make some strong tea, then dissolve (as best you can) a small bit of an iron supplement pill in water or vinegar. Add the iron solution to the tea and watch the effect. For the commercial fabrication of ink, gallnuts were used.

and red wine. Always plenty of the black stuff to write poems and sign death sentences with. But the pencil, that is a different story. Far from being just for children, it was, and is, an essential tool for artists, engineers, carpenters, and architects.[72] At engineering school in the late 1980s we still made (some of us did anyway) beautifully crafted pencil drawings of double-mantled stainless steel reactors. And in the army, close to the polar circle four years earlier, did we write out orders and decipher incoming radio messages with ballpoint pens? We certainly did not—in fact, this was forbidden because the ink in a pen may easily freeze.

The 'lead' in the pencil (which is obviously not lead as in the element 82, but something else) brings us to these green valleys of the Lake District and Cumbria, England—as unlikely a place for an information technology hub as the orange orchards around Palo Alto. The different is that in California in the 1970s it was the dedicated people that mattered, not any local silicon mines. In Borrowdale in the late sixteenth century, it was the inside of the mountain itself that made the difference, for there you find the stuff from which to make pencil lead.

Not that the people were unimportant. Entrepreneurship thrived in different forms. 'Black Sal', for example, working out of the small town of Keswick close to Borrowdale, was allegedly running a pencil-lead smuggling network in the early eighteenth century. The precious cargo was hauled over the rough hill-land down to the Irish Sea, where waiting boats could take the *plumbago* to the continent. Clandestine mining that fed the smuggling business was the order of the day, and one or two armed robberies of the mines were also recorded. All in all, it seems rather like the

FIGURE 20 A piece of Borrowdale graphite bought, quite legally, in Keswick Cumbria, UK, 2012. Photo by the author.

Wild West, with redcoats replacing the blue US cavalry, and the locals trying to hang on to what they saw as their property against 'foreign' owners and investors.[73]

In those days many kinds of mining were important in Cumbria,* but the black stone called *wad* or plumbago was the most precious mineral ever dug up from the ground there. Its first use is supposed to have been in late medieval times for keeping track of whose sheep belonged to whom—an important enough part of IT then as it is now; accounting for our earthly possessions.[74]

So what is this grey stuff if it is not lead? It is a crystalline form of carbon called graphite (shown in Figure 20), very different from another form of crystalline carbon, diamond. But just as in a diamond mine, at Borrowdale workers leaving the site were searched for hidden stones, and armed guards followed the graphite transports to their final destinations.

* Or rather Cumberland, the smaller district that has been part of Cumbria, a UK county, since 1974.

These different forms of carbon are a bit confusing. You mine carbon in a coalmine as well, and in Swedish it is even more confusing as the words 'carbon' and 'coal' are both the same, 'kol'.* And then there are the charcoal kilns where you supposedly make coal out of wood. There is no confusion here—it is the same element we are talking about, and not a case of mistaken identity as with the 'lead' in the pencils.

So let's start with the most precious member of this family: diamond. In this material every carbon atom is attached by a strong bond to four other carbon atoms. This forms (if we could see it) something like a miniature scaffold, but one in which the spokes do not make 90° angles with each other but are separated by 109.5°, enabling this network to expand forever in all directions. As the spokes are made of strong carbon–carbon single bonds, this is a very tough structure. These carbon atoms are also completely unimaginative: they arrange themselves in exactly the same way in the whole diamond crystal, which is why diamond is a crystal in the first place— we only need to know the position of one single carbon atom, and then we can tell where all the other carbon atoms are in the crystal.

Furthermore, the electrons are kept under tight discipline, one pair of them for every carbon–carbon single bond, guarded by the attractive spell put on them by the positively charged carbon nuclei. This means that diamond should be an electrical insulator and not conduct electrons—and so it is, and a very good one at that. (Surprisingly, it is a good thermal conductor, as we saw in Chapter 8.) The electrons are held so tightly that they cannot be moved by visible light either, in the way that materials normally obtain colour, making a perfect diamond completely transparent.

---

* And to make matters worse for us, 'kol' is a homonym of cabbage, 'kål'.

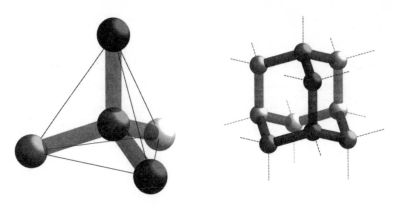

FIGURE 21 Left: a carbon atom attached to four other carbon atoms forming what we call a tetrahedron. Right: many carbon tetrahedrons binding to each other forming the diamond structure. A diamond is an infinite network of carbon atoms, much like one big molecule, and the structure continues along the dashed lines.

Moving down the price list to graphite, we now see the carbon atoms arranged in a completely different way, which is the very definition of what are called *polymorphs*. These are materials with the same chemical composition but different three dimensional arrangements of atoms. In the special case of a pure element appearing in different disguises, they are known as *allotropes*. In graphite, instead of the three dimensional network in diamond, the carbon atoms form honeycomb-shaped sheets one atom thick, with every carbon atom now forming bonds to only three other neighbours. The most eye-catching feature of this structure is the completely symmetrical hexagons formed by six carbon atoms in a closed circuit.

If you are an accountant, you may have immediately spotted that we now have more electrons than we need in order to put a pair of electrons between each carbon atom, thus producing the chemical bonds of the hexagonal layers. Each carbon atom in

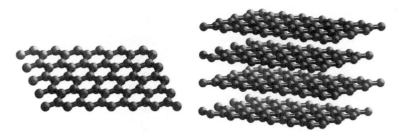

FIGURE 22 A small part of a graphite sheet and the stacked sheets in a piece of crystalline graphite, as from the Borrowdale mines.

diamond contributed four electrons, one each to every neighbouring bond. The same exercise for graphite leaves us with one surplus electron for each carbon, after distributing one to each of the three neighbours.

What to do with these extra electrons? A natural thought would be to use these to bind the sheets together to form the three dimensional material that we pick up as a real piece of graphite, but this is not what happens. There is nothing resembling chemical bonds between the layers of graphite. Instead, the closeness of the carbon atoms in the layers creates a space on top of and under the sheets, where these extra electrons are free to move while contributing to the bonds in the hexagons. These are not double bonds as we saw in Chapter 5—we call them instead *delocalized double bonds*, because we cannot pinpoint exactly which carbon atoms the extra electrons are between. Freewheeling electrons like these mean electrical conductivity, and graphite is indeed an excellent conductor of electricity—as long as we stick to one layer that is.

The extra bonding also means that the sheets are strong, and only very reluctantly are these extra electrons contributing to holding two sheets together—the distance between the layers is more than double the C–C distance in the layers.

Another important property is that the electrons can now also be moved by visible light, although my colleagues may be upset by this terminology and would probably prefer me to say that the electrons are 'excited'. As the electrons are constantly moving, or behaving more like a density wave (an indefinite cloud surrounding the atomic nuclei), maybe we should instead say that they shift gear. Chemists frequently use the term 'jumping' for exciting electrons, especially when they graphically visualize this with diagrams on paper. For most substances that can absorb visible light there will be only one, or a selected few, possible gearshifts, each requiring its own colour, or wavelength, of light. These colours will, so to speak, be absorbed by the material, and taken out of the complete spectrum of white light, leaving us to see those colours that remain.

But graphite is black, or dark grey with a metallic lustre, so it needs to absorb all the wavelengths of the incoming photons. Therefore it cannot have a normal gearbox with five or six positions, it needs continuous gear shifting from the low-energy red colours to the high-energy shades bordering on the ultraviolet. This is exactly what happens when these extra electrons that are not confined to the C–C single bonds in graphite can travel freely in what is really one gigantic, flat, molecule.

Finally, the flatness will have the last say in why graphite became such a vital part of information technology. As there are no strong bonds holding the sheets together, it is relatively easy to make new, more robust, attractive forces by pressing the graphite against a rough surface. Then the sheets will stick to the fibres of the paper, and large chunks are rubbed off, filling the microscopic valleys of the paper. These are not the single layers of atoms we see in a pencil drawing—they are hundreds to several thousands of atom layers thick.

Having said this, it should be noted that single atom layers of graphite can be lifted from a graphite surface, and this is the thinnest of all known materials, called *graphene*.* Work on this substance was awarded the Nobel Prize in Physics 2010. It is not altogether unlikely that this material will also find its way into information technology sometime in the future.

Further down in the value-per-dig ladder we find something called anthracite, but now we are leaving the pure substances and coming across materials that are really mixtures. Anthracite may be up to 97 per cent carbon, but because of the impurities it lacks the long-range order of graphite, although it will contain some very small crystalline particles as well.

Further down the quality value chain, materials like coal are in fact rather complicated to characterize properly in terms of their molecular content, but it is clear that with a lower carbon content the crystallinity goes down. This is just as you might expect from observing samples of table salt in your kitchen: the very pure, industrially made, NaCl forms uniform cubic crystals that are nice and transparent—under the microscope they look like little gemstones. The large variety of other 'gourmet' salts, sea salts, and other preparations on the market usually have a much higher content of impurities and are consequently less crystalline.

There were graphite mines in many places in Europe in the sixteenth to eighteenth centuries, and also farther afield, but the graphite mines in Cumbria stood out for the extraordinary quality of the graphite. It was very pure and highly crystalline—in fact, it was the world's only known source of graphite that would make

---

* There are also other forms of carbon, the famous C_{60} molecules, the diamond sibling lonsdaleite, and several more.

decent pencils. For a while it was also important for making moulds for cannon balls, but here lower-quality graphite could also be used, and replacement technology was soon developed.

When the Swedish industrial spy and gentleman Reinhold Angerstein (Chapter 4) visited Borrowdale in 1754 he duly noted the calamities of an earlier generation, but by that time the authorities and big industry seemed to have got the situation under control.[75] The heyday of Cumberland graphite was soon to be over though. In another generation, the French Revolution and the British export embargo to the newly formed Republic had the surprising and unwanted (for the British) effect of the invention of the modern pencil by Nicolas-Jacques Conté. He combined clay and substandard graphite, baked the mixture, and enclosed it in cedar tree sticks—all described in the French patent number 32.[76]

The last Borrowdale mine was closed in 1891 as there was no more minable graphite to be found. The pencil company lives on however, as the Derwent Cumberland Pencil Company, although it is no longer an independent enterprise. Some of the late eighteenth-century pencil-making companies, such as the ones started by Conté in France and Kaspar Faber in Germany, still live on, showing that this is indeed still a vital part of our information technology.

In the novel *Hand of Glory* by Glen Petrie,[77] the notorious smuggler Black Sal is hunted down and killed by wolfhounds, a grim fate indeed. Hugh Walpole does not describe life around Keswick in gay colours either, in the more well-known novel *Rouge Herries*,[78] although ill deeds explicitly based on pencil-quality graphite are absent from that work. But as affirmed by Angerstein, things were calming down, and in 1807 William Wordsworth published the poem *I Wandered Lonely as a Cloud,* with references to daffodils and the beauty of the landscape rather than to mining and crime. Whether the first draft of this poem was jotted down in pen or pencil is not known.

You can do many things with graphite besides making pencils. It is a great lubricant, but also, because it conducts electricity, it makes good electrodes for use in large-scale industrial applications such as in the production of aluminium metal from aluminium oxide, which, as it happens, is an underlying theme of the next chapter.

10

The Emperor and Miss Smilla

*In this chapter an aeroplane crashes into a bog, an emperor meddles
in chemistry, and a young lady specializing in the crystalline form of
water will guide us to the chemistry of aluminium and fluorine.*

Biking south from Avignon, brief residence of the popes in
southern France, towards Arles, town of bulls and van Gogh,
should be a leisurely experience in the smiling Provencal landscape, making no undue demands on one's physical abilities. That
is, providing you stay away from the only obstacle on the way—
the fortified hilltop village of Les Baux-de-Provence.

An odd (some would say suspect) error of navigation brought
us to the top one sunny day in September many years ago, but this
detour proved to be well worth the effort. Both the village and the
view are spectacular, and well justified claims to fame for Les
Baux. However, Les Baux is, or should be, famous for another
thing, the ore known as *bauxite*, discovered in the vicinity of the
village by Pierre Berthier in 1821.

Through the bauxite ore there is a curious connection between
this sunny place of cicadas and *pastis* in the afternoon shade,

and the Greenlandic adventures of Smilla Qaaviqaaq Jaspersen. These adventures that, told in the novel *Miss Smilla's Feeling for Snow*, propelled Danish author Peter Høeg to international fame and fortune in 1992.[79]

In this bestselling novel, that can be described both as a thriller and as a 'post-feminist' critique of Danish colonialism,[80] the mysterious doings of the Cryolite Company of Greenland plays a major role, as does Smilla's profound knowledge of the solid-state properties of water. A former glaciologist of Danish–Greenlandic origins, she investigates the death of a neighbour's six-year-old child after a fall from a snow-covered roof, which is dismissed by the police as an accident. She finds herself rummaging in the archives of the Cryolite Company in Copenhagen, examining forensic evidence, and finally joining the crew of an ill-fated expedition to a remote part of Greenland.

At first sight, the common denominator between bauxite and *cryolite* is aluminium. As such, this is not so remarkable: aluminium, right under boron in the Periodic Table, is present in many minerals and ores, and is the third most abundant element in the earth's crust. However, the important connection joining these materials is actually the process of obtaining aluminium metal. The metal proved to be an enigma to nineteenth-century chemists, and seemed to be proof of the Swedish proverb 'when it rains soup the poor man has no spoon'. It was everywhere, yet impossible to get hold of. Bauxite was clearly a good starting material. It contains different forms of aluminium hydroxides like $Al(OH)_3$ and $AlO(OH)$, and these can be converted relatively easily to aluminium oxide (Al_2O_3). But how do you separate the aluminium ions from the oxide ions and get them into the metallic state?

A nephew of Napoléon Bonaparte, also a self-styled emperor of the French under the name of Napoléon III, predicted the future importance of the metal for arms manufacture, and his instigation eventually lead to the Hall–Héroult process patented in 1886 and still in use today.*

The story of how Charles Hall in Ohio, US, and Paul Héroult in Normandy, France, independently transformed the metal that was once used for the cutlery of the most important guests at the court of Napoléon III to the metal we use for soft-drink cans today is fascinating, at least for a chemist, but has been told many times.[81]

We will, however, dwell a little on the method itself, as it demonstrates some important principles of chemistry, and it explains such disparate facts as why there are aluminium plants on Iceland but no bauxite mines, and why a B17 Flying Fortress ended up in a bog in western Sweden on 24 July 1943.

The Hall–Héroult process uses electrolysis. You may recall how we needed a really good reduction reagent to make uranium metal from uranium 4+ ions in Chapter 2. Well, to make aluminium from Al_2O_3 is even harder, and instead of finding a chemical reagent that can provide the necessary three electrons to make the Al^{3+} ion into a neutral metal, we turn to naked electrons.

In electrolysis, we introduce electrons in one electrode (a conducting stick) to a solution of the Al^{3+} ions. At this electrode, the electrons jump out into the solution to be caught by the Al^{3+} ions. When the Al^{3+} ions have gobbled up three electrons in this way we will get Al(0) atoms, and these will rapidly combine to form aluminium metal that falls to the bottom of the reaction vessel. On the

---

* Oddly, Napoléon III has another entry in chemistry history as the initiator of the invention of margarine.

other side, a second electrode is waiting to take up the electrons injected from the first electrode in order to close the electric circuit. As no electrons are let through from the inlet electrode these need to be taken from somewhere else, and if the solution you have is molten Al_2O_3, then there are only the O^{2-} ions left. These are negatively charged, so it would make sense that they travel to a positively charged electrode and give away two electrons.

$$Al^{3+} + 3e^- \rightarrow Al$$
$$2O^{2-} \rightarrow O_2 + 4e^-$$

As no electrons are allowed to escape, the number taken up by the aluminium must exactly match the number given away by the oxygen, which gives us the overall reaction:

$$4Al^{3+} + 6O^{2-} \rightarrow 4Al + 3O_2$$

This is of course the exact opposite of the reaction that would occur spontaneously, if the oxygen molecules could only crack the thin but impenetrable oxide sheet protecting the surface of all aluminium objects, which gives them fantastic anti-corrosion properties. The reason we can do the exact opposite with electrolysis is that we put a lot of electrical energy into the reaction, which is transformed into chemical energy stored up in the compounds formed. This is a very costly reaction. If you want to do the schoolbook example of electrolysing water you can use a normal battery, but to make aluminium you had better get yourself a hydroelectric power plant or some other source of cheap electricity. This explains why Iceland has an aluminium industry. Bauxite is very inexpensive and can be transported by boat, but few places can provide electricity at such a bargain price as Iceland.

At the time of Napoléon III, all this was known. The problem was that the high melting point of aluminium oxide, 2,072°C, made the process all but impossible. Then along came Hall and Héroult, who showed that you could dissolve aluminium oxide in melted sodium hexafluoroaluminate (Na_3AlF_6), reducing the temperature needed for the process by about 1,000°C. The snag was that Na_3AlF_6, better known as cryolite, could be found only in one place on earth, namely in the Ivittuut mine in southern Greenland. Consequently, when aluminium really became important for the military, the Greenland cryolite mines became a vital strategic asset, as Høeg's heroine Smilla discovers in the Cryolite Company's archives in Copenhagen.

The occupation of Denmark by Germany in 1940 made the British and its allies nervous. Under the cover of a journey to the Northwest Passage, the only Canadian government vessel able to navigate the icy Greenlandic waters, the *St. Roch*, under the command of Henry Larsen of the Royal Canadian Mounted Police, set out from Vancouver to survey the situation as there was fear of a German invasion.[82] Later on, with the entry of the United States into the war, the cryolite question was resolved by Greenland temporarily becoming a US protectorate, and the production of the Ivittuut mine increased substantially.[83]

In his novel Peter Høeg hints that there was a German plan to attack and occupy the cryolite mine during World War II. The only recorded Nazi attempt on Greenland was, however, a modest effort to establish a weather station. The humble invasion force of 17 men was, however, soon discovered and dealt with by the Danish (Hound) Sledge Patrol and the US Army Air Force.*[84]

---

* Under the command of the legendary Norwegian polar aviator and USAAF colonel Bernt Balchen.

Instead, the Germans set up a factory for producing synthetic cryolite next to the aluminium plant in Herøya in southern Norway. This process was rather new at the time, but Nordische Aluminium never saw full-scale production as it was the target of a successful bombing mission. Not only were the factories destroyed: of the 180 B17s dispatched in the morning of 24 July 1943, only one was lost.[85] However, through skilful navigation and piloting by first pilot Osce Vernon Jones, the damaged aircraft, a B17 called *Georgia Rebel*, landed safely on neutral ground: a bog in western Sweden close to the small town of Årjäng. This was the first of over 200 such US Army Air Force emergency landings in Sweden during World War II.[86]

At the time when Høeg wrote his novel, the cryolite mines had been exhausted for a couple of years, and man-made cryolite was being used in the aluminium process. Thus, while Smilla's adventures spin off in another, more biological direction, we take up the cryolite trail in a recent thriller by Swedish physicist and venture capitalist Lennart Ramberg, *Kyoto and the Butterflies (Kyoto och Fjärilarna,* 2007).[87] This novel also takes us on an expedition to the extreme north, but this time the goal is stated from the beginning: to find the mysterious source of the tetrafluoromethane molecules detected over the Arctic by an eccentric chemistry professor gone missing. It is styled as an ecological thriller, since the tetrafluoromethane molecules produce a greenhouse effect more than a thousand times stronger than carbon dioxide.[88] The novel also takes some pains to describe in detail such favourite instruments of the analytical chemist as the gas chromatograph, perhaps for the first time in popular literature.

The main character, the PhD student Kimi, is rather unsure about himself. This seems partly due to his former career as a model, and partly (although quite reasonably) due to his professor going up in smoke, and with him Kimi's hope of completing his degree. However, he does know all about tetrafluoromethane (CF_4), and the environmental campaigning group he sets up camp, or rather ship, with, knows all about publicity.

The villain of the story makes cheap aluminium close to the polar circle, as cryolite also, of course, plays an important role in this novel. The connection here is the 'functional' part of the mineral, its fluoride ions. These are responsible for the ability of molten cryolite to dissolve the aluminium oxide, but they also have a disturbing tendency to combine with carbon atoms from the graphite electrodes used in the process to form CF_4, especially if the aluminium smelter is run on a shoestring to maximize short-term profits.

As befits a 'global warming' suspense novel, carbon dioxide is not ignored, but plays a short and rather unexpected role in its solid form, also known as 'dry ice'.

The two novels are widely different, apart from some thematic similarities. Ramberg does, however, cite Høeg's novel as a major inspiration to writing in general. The impact on the authors' lives also differ, as Ramberg's novel is yet to be translated into English, whereas one presumes that the huge success of *Miss Smilla's Feeling for Snow*, or *Smilla's Sense of Snow* as it is known in the US, and the international movie version (directed by Billie August, with Julia Ormond as Smilla), has made life financially comfortable for Peter Høeg. As for the two inventors of the Hall–Héroult process, they both made a fortune, but Napoleon III did not live long enough to see the project realized.

Osce Vernon Jones was repatriated from Sweden late in 1943, resumed service at the Ridgewell airbase in Essex, home of the

381st Bomb Group, in January 1944, was promoted to major, and survived the war. He died in 1989. The 'real' Danish cryolite company that processed the cryolite from Greenland was called 'Øresund's chemiske Fabriker A/S', and in 1987 the last commercial cryolite was shipped out of from Ivittuut.[89]

11

Rendezvous on the High Plateau

In Chapter 11 we are confused by a red, blue, and white flag, go to the movies with Gordon Gekko's father, and learn more about electrolysis, nuclear reactions, and isotopes.

The two men in white anoraks were slowly approaching, skiing in the bitter cold over the Hardangervidda mountain plateau in the winter of 1943. Were they friends or foes? This was a matter of life and death for the six young men watching the only other living beings in sight for miles of snow-clad wilderness. Their pace was slow, the men were thin and didn't look too well, just as if they might well have spent 130 days of the winter of 1942–43 hidden in a rudimentary hut on the mountain, surviving on moss and poached reindeer. It had to be them. The group's leader, Joachim Rønneberg, decided to make contact.[90]

This story is first a tragedy and then a success, and it does not begin on the Hardangervidda but in Scotland where Britain's ski capital, the small town of Aviemore in the Cairngorms National Park, is going to be our starting point for several dangerous journeys across the North Sea.

A few years ago we drove up the main mountain road, eventually leading to the Cairn Gorm peak itself, 4,084 feet (1,245 metres) above sea level, and passed the park's visitors' centre located in pretty surroundings by a small lake. We glimpsed something flapping in the wind that did look a bit like the Union Jack, an unlikely occurrence in the highlands.* We turned around and took the path up the hill, and soon discovered that what we first mistook for the British ensign, because of its colours, was in fact the Norwegian national flag.

In 1468, when the Norwegians gave away their last Scottish possessions to King James III in Edinburgh, the Norwegian flag had not even been invented, so we were a wee bit curious as to why it was flying here, in the heart of the Cairngorms.

But of course, mountains, snow, and skiing—what could be more Norwegian? And this simplistic reasoning is actually closer to the answer than we might have thought, as a commemorative sign told us that on this spot were the lodgings of the famous Kompani Linge during World War II. These Norwegian commandos operating behind enemy lines in their occupied homeland, at one time led by Captain Martin Linge, needed training grounds as closely resembling Norway as possible. Also, being part of the Special Operations Executive (SOE), the secret armed branch of the British Government, rather than the regular army, meant that a remote location was an additional benefit.[91]

Many military campaigns have been conducted for the sake of gold and silver, carbon and oil, but the most famous operation of the Kompani Linge was planned for the sake of water. Not normal water, H_2O (sadly, these wars will surely come), but its chemically

---

* Note that this was written before the 2014 referendum.

FIGURE 23 Kompani Linge memorial outside Glenmore Forest Park visitors centre, Loch Morlich, Scotland, UK. Photo by the author.

near identical twin, a compound called di-deuterium oxide. In this molecule—also known as heavy water, or by its formula D_2O, where 'D' is for deuterium—the isotope of hydrogen that is composed of one electron and one proton, just like normal hydrogen, but with an additional neutron in the nucleus.

Heavy water sounds a bit sinister—we think of 'heavy metals', the bad boys of the Periodic Table—and you may have some vague notion that D_2O might have something to do with nuclear power or atomic bombs. This is understandable, and I hope I will not shock you, but you drink heavy water every day without any harmful effects whatsoever. And, come to think of it, some of the bad boys are quite friendly too and essential to keep you alive.

When we talk about an element of the Periodic Table and use its symbols, what we really mean, even though we might not realize it, is the natural occurring mixture of its isotopes. We have already encountered the two naturally occurring boron isotopes for example, and a tablespoon of silver will contain 51.8 per cent of the silver isotope with 47 protons, 47 electrons, and 59 neutrons, and 48.2 per cent of the isotope with 47 protons, 47 electrons, and 61 neutrons.

Chemically such pairs of isotopes are almost identical twins, because what matters in chemistry is the number of electrons, which the atoms use to make chemical bonds, and the number of protons, the number of which determine how firmly the negative electrons are held by the nucleus. The obviously neutral neutrons do not meddle in this game as they neither attract nor repel the electrons, and they have no influence on the strength of the chemical bonds formed by the electrons. This was the idea behind *Operation Spanner* in Chapter 2: normal chemical analysis would not have picked up that the isotope ratios had been fiddled with.

So, normally when we talk of water we talk of H_2O, where the 'H' now has to represent both these stable isotopes, and where they occur with the proportions 98.98 per cent normal hydrogen and 0.02 per cent deuterium, the so-called natural isotope ratio for hydrogen. The oxygen atom does not care whether it binds to an H or a D, and a water molecule may contain none, one, or (very rarely) two atoms of deuterium. This means our drinking water contains H_2O, HDO, and D_2O in the approximate proportions 25 million to 5,000 to one. The chances of meeting a D_2O molecule thus seem slim at first glance, but as there are about 500,000,000,000,000,000,000,000 water molecules ($5 \cdot 10^{23}$) in a tablespoon of water, you can make a safe bet at your bookmaker's that you will consume a fair number of heavy water molecules each day.

So we eat and drink it, but can you make it in pure form? The answer is yes, and exactly because of this you should not replace your daily intake of H_2O with D_2O. The chemical bonds formed by the deuterium to carbon, nitrogen, and oxygen in the body will be close to exactly the same as for H, but the speed with which our enzymes (the protein catalysts in our body) shuffle H or D around between molecules will be slightly different, because D weighs twice as much as H. This means that a diet of heavy water will slowly distort your metabolism, and in the end give you severe health problems. This slight difference in speed—or rather reaction rate, as chemists like to call it—is the basis for the production of pure heavy water.

As the neutron was only discovered in 1932, and the deuterium isotope shortly afterwards, it was not surprising that the first reports on Nazi interest in heavy water were met with some scepticism in intelligence circles, mostly devoid of officers with science education.* Luckily, physicist Reginald Jones had been recruited as the first scientist to the intelligence services in 1939, and could act immediately when he received a telegram from a Norwegian scientist informing about the Nazi plans to increase the heavy water production in occupied Norway.[92]

Although in high concentrations and over a prolonged period heavy water is not good for you, the German plan was not to slowly poison the British with D_2O from Norsk Hydro's Vemork facility in Rjukan. Although not a prime reason for the occupation of Norway in 1940, getting their hands on the world's only large-scale facility for heavy water production was sure to profit the Nazi regime's atomic bomb project, the Uranverein.

---

* Note for example that the 'Cambridge Spies' Maclean, Philby, Burgess, and Blunt had degrees in modern languages, economics, history, and art history respectively.

Unfortunately for the Nazis, the Rjukan site, in the south Norwegian mountains in Telemark County, was no secret to the Allies. It was part of a large assembly of chemical plants where the main reagents were electrons—very cheap electrons, used in a variety of processes. One of these was hydrogen gas production by the electrolysis of water, giving D_2O-enriched water as a by-product. Heavy water had been produced here since the mid-1930s, the last shipment before the war being clandestinely obtained by the French Deuxième Bureau, the military intelligence, and the 185 kg hurriedly brought to safety in England by two French scientists fleeing the German invasion in the early summer of 1940.[91]

The remote site was a mixed blessing: easy to defend and patrol, in many ways it was also more vulnerable to daring commando operations than if it had been located in, for example, Ludwigshafen on the Rhine. Ludwigshafen, although home to a large

FIGURE 24 The Vemork power plant in Rjukan. The Gundersen Collection/ Norwegian Industrial Workers Museum.

chemical conglomerate, lacked the essential resources of Rjukan: the cheap electrons you get from hydroelectric power.

If you have electricity, you can perform electrolysis by passing current through a solution using two electrodes, as we saw in Chapter 10. If the solution is water, and the voltage is high enough (this is a classic school demonstration experiment), you will get hydrogen gas where the electrons enter the solution and the cations assemble (the cathode electrode), and oxygen where they are taken up again (the anode electrode) by the closed electric circuit. If the solution is a melt of aluminium oxide, you can produce metallic aluminium at the cathode, and from a concentrated solution of sodium chloride (table salt) you can manufacture chlorine gas at the anode. All these processes are workhorses of the worldwide chemical industry.

In Rjukan, the prime interest of Norsk Hydro was to make ammonia to be shipped further to one of their factories in Herøya (yes, the same place as in Chapter 10), where it was converted into nitric acid and then into fertilizers. Originally, nitric acid was made directly in Rjukan using an electric arc method that is very costly in energy—making nitrogen and oxygen in the air react directly— but by the 1930s this had been replaced by the much more efficient Haber–Bosch process. The Haber–Bosch reaction combines nitrogen from the air with hydrogen gas to form ammonia:

$$N_2 + 3H_2 \rightarrow 3NH_2$$

This does not need an electric current, but Hydro instead used their electric power to make the hydrogen gas needed by electrolysing water.*

---

* Today natural gas, methane for example, and other fossil resources are used to make the hydrogen gas that is fed into the Haber–Bosch process, making the majority of the nitrogen atoms we have in our bodies dependent on non-renewable resources.

It was soon realized by, among others, the young inorganic chemistry professor Leif Tronstad at the Norwegian Institute of Technology in Trondheim, that this process enriches D_2O and HDO in the water that has not been converted to gases, because H is less heavy than D and the reaction producing H_2 will be faster than that giving D_2.[93, 94] You can imagine this as the smaller H^+ ions speeding more quickly to the negative cathode, where they can pick up an electron and form H_2 gas by combining with another sprinting H^+ ion, leaving the fatter D^+ ions slightly behind in the race. The H atoms will therefore be removed from the solution faster than the D atoms, leaving the water left behind slightly enriched in D_2O and HDO.

The difference in reaction speed is small though, so your one-pot water and battery-based school demonstration would have been of little value to the Germans. Tronstad and others instead designed a process where this first solution is passed on to a second electrolysis cell, where it again gets slightly enriched, and so on in a so-called cascade process until the umpteenth cell, where you finally get your highly enriched heavy water.

In doing this you have wasted an enormous amount of energy. Not only have you electrolysed all the H_2O molecules, but a large number of the HDO and D_2O molecules too. To do this on an industrial scale you need a huge power plant, and the Vemork had been the largest hydroelectric power plant in the world when it was first operational in 1911. Electrolysis uses direct current (DC)—this is what you get from a battery, as opposed to the alternating current (AC) that we use in normal electrical appliances at home. In theory it is better to transport electric power by direct current, but all the engineering problems had not been sorted out at the time, so in the 1930s and 1940s you needed to be close to the power plant to operate a good-sized electrolysis factory.

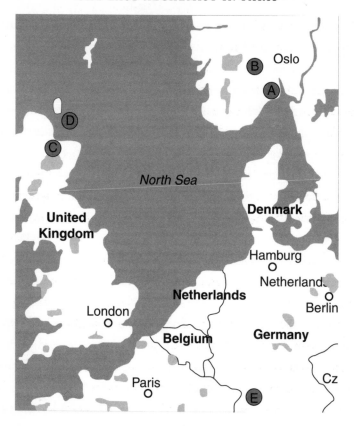

FIGURE 25 A: Herøya, Norway, Norsk Hydro site; B: Rjukan, Norway; C: Aviemore, UK, SOE base; D: Wick, UK, RAF base; E: Haigerloch, Germany, experimental reactor.

Tronstad had escaped to London in 1941, and with both a military and scientific background soon became employed by the exiled Norwegian High Command. As he had an intimate knowledge of heavy water production, in his diaries invariably called 'squash', he would play a major role in the various operations collectively known as the 'heavy water sabotage'. Prior to departing, it was he who sent the first important 'heavy water' telegram to Reginald Jones and subsequently parted with more vital intelligence, although on condition that no information was passed on to Imperial Chemical Industries (ICI), the UK chemical giant, as 'blood is thicker even than heavy water'.[92]

The first of the military actions, Operation Grouse, successfully parachuted four Norwegians from Kompani Linge onto the Hardangervidda in late 1942 for reconnaissance work. The second was to become a tragedy.

Two gliders drawn by two Halifax bombers with a total of 48 young, specially trained volunteers from the Royal Engineering Corps, took off from the RAF Skitten base outside Wick in north-eastern Scotland on the evening of 19 November 1942. Only one of the bombers came back, and most other things went wrong in Operation Freshman. It was impossible to find the right landing spot, the weather was bad, one Halifax crashed killing the crew of seven, and the gliders made difficult landings far from the target leaving some of the soldiers wounded.

The Nazis, soon alerted by the local authorities (who did not have much choice in the matter and should not perhaps not be judged to harshly), brutally executed all the survivors, as they had civilian clothes under their uniforms and therefore were considered to be partisans, not regular soldiers. In retrospect it was a badly planned operation. After the action the soldiers were supposed to make it over the mountains into Sweden, but they did not know how to ski and knew only a few phrases of Norwegian. Their commanding officer, Colonel Henniker, was not happy with the plan and probably sensed he was sending the young men to an almost certain death.[95]

In the run-up to the operation the soldiers were told that a success could eliminate a threat that could possibly turn the war and make the Axis powers victorious within six months. This was, as we now know, a huge exaggeration, but at the time there was no knowing that Heisenberg and his team were making such slow progress.

The heavy water itself is harmless—you cannot make an atom bomb with that—but it has a vital role in a nuclear reactor based on non-enriched uranium, the only type envisaged at the time. If a uranium rod is hit with neutrons directly released from a newly split uranium atom, these particles will have such high speed that they will not have enough time to react with the uranium-235 nucleus, cleave it, generate energy, and propagate the chain reaction, or to combine with the much more abundant uranium-238 isotopes to give, after ejection of an electron and some other radiation, the sought after plutonium-239 isotope from which you can make an atomic bomb.

These neutrons are known as 'fast neutrons' and need to be slowed down by collisions with other particles, which is best done by making them collide with something of their own size, like a proton. You can imagine a set of pool balls moving at full speed—putting a few full-size bowling balls in their way will merely make them bounce and alter direction, but not speed. In making them collide with other pool balls, the speed drops and the directions change, as anybody who has ever approached a billiard table will know.

The problem with the protons in ordinary water is that instead of bouncing the neutron off and slowing it down by taking up some of its speed, it may absorb them. In such a nuclear reaction the completely stable isotope deuterium will be formed, efficiently stopping the reaction. On the other hand, deuterium is not as good at munching up the neutrons, and to make a nuclear reactor in 1942 there were essentially two choices of such *moderators*: graphite—that is, pure carbon (with no hydrogen, and small carbon atoms that do not munch neutrons)—and heavy water. Heisenberg's team had no luck with graphite,* although the Americans were successful with this material, and thus concentrated on D_2O. And they needed a lot of it.

Hence the global interest in the drop-by-drop formation of tons of heavy water in the village of Rjukan. This dripping is a reoccurring theme in the 1965 movie *The Heroes of Telemark*, because of course the British did not give up. However, the four men from Operation Grouse had to endure almost the whole winter of 1942–43 without additional supplies in their hut on the Hardangervidda, 3,000 feet (1,100 metres) above sea level.

Of course, the movie does not faithfully portray the real events of Operation Gunnerside, possibly the most successful behind-the-lines sabotage during the war. There is too much to be told in too short a time, plus the need to add on some kind of love story. However, the film comes much closer to the truth than *Operation Crossbow*, about the countermeasures against Hitler's V1 and V2 rockets released the same year.

The Oslo physics(!) professor, played by Kirk Douglas, vaguely resembles James Bond—which is made clear in his unconventional photographic darkroom collaboration with a young female assistant in the film's opening scenes—and has no real life counterpart, but seems to be a mixture of chemistry professor (and Major) Tronstad and commando leader Joachim Rønneberg. At this stage the Norwegians and the SOE judged Tronstad, against his wishes, to be too important to be risked behind enemy lines.

---

* Carbon has a stable isotope with mass number 13, but as normal carbon is C-12, with an even number of protons and neutrons, this is a very stable and unreactive nucleus that will not react with neutrons and make C-13 to any great extent.

But the essentials are correct. Norwegian irregular troops from the SOE, under the command of Rønneberg, were parachuted in during the early months of 1943, making contact with the Grouse group and sabotaging the electrolysis cells in the Rjukan factory. Half the group escaped to Sweden, and the other half stayed behind and later sank a rail ferry transporting heavy water due for the experimental nuclear reactor in Haigerloch, Germany, using time-set bombs. This, and a US bomb-raid, put an end to the German effort to obtain heavy water from Rjukan.

In the end, this did not win the war, but there was no way to know this at the time. Surely it was an immense boost for morale, both for the British and the Norwegians, and one less thing to worry about for the war cabinet, because these questions inevitably reached the highest levels of attention. As Reginald Jones, later professor in Aberdeen, put it: 'the price was just too high' if the Germans had succeeded.

The 'Grouse' and 'Gunnerside' commandos all survived the war, and the dead heroes of 'Freshman' have not been forgotten either, with memorials both in Scotland and in Norway.

In the final phase of the war, the Norwegian exile government was concerned that the Germans would destroy vital parts of Norwegian industrial infrastructure, especially the hydropower installations in Telemark. This gave Professor, or rather Major, Tronstad the chance of active service at last. He had the same training as the other, often much younger, soldiers of the Norwegian Independent Company 1, as they were officially known, and had been eager to participate ever since arriving in the UK. In November 1944, he and eight other Linge commandos were parachuted in over Hardangervidda, and Tronstad took command of

Operation Sunshine. By early spring 1945 he had over 2,000 armed resistance fighters under his command.

On 11 March 1945, just a little less than two months before the German capitulation in Norway, Leif Tronstad was killed in a man-to-man fight with Norwegian collaborators. Not far from the hut in Syrebekkstølen in Telemark where he died, a memorial stone has been raised, commemorating the life and death of Tronstad and fellow resistance fighter Gunnar Syverstad.

Heavy water was produced in Vemork until the end of the 1960s when the hydrogen plant was closed down, and the entire site was closed by Norsk Hydro in 1991. It now houses the Norwegian Industrial Workers Museum.

12

The Last Alchemist in Paris

*In this chapter we learn about heavyweights and lightweights, play
with atomic balls that are soft or hard, big or small, some which may
keep you sane, or, on a bad day, completely block that genial burst
of creativity.*

The Periodic Table is packed with elements. This is hardly
surprising, but it is also full of numbers: atomic numbers obviously, the number of protons in the nuclei, that form the address
of an element in the table, but also atomic weights, number of
isotopes, and any other property directly linked to the atom in
question. In short, if you are inclined towards numerology and
proving your favourite theory by combining two or more numbers
to give a pleasing figure or neat coincidence, you have more opportunities here than with all the pyramids of Egypt combined.

For example, gold (Au) has 79 protons and so consequently
atomic number 79. Yttrium (Y), the first of seven elements to be
found and named after the mine in Ytterby on Resarö Island in
the Stockholm archipelago, has atomic number 39. One of the later
discoveries from the same mine was scandium, atomic number 21,
in 1879. Add these two numbers and you get 60, a common enough

street number, but of special significance in this chapter. Divide 39 by 13, the number of bad luck, and you get 3, the atomic number of lithium (Li).

In February 1896, a curious man in something of a state, and not touched by good luck for some time, checked in to the Hotel Orfila, on rue d'Assas 60, close to the Jardin du Luxembourg in Paris. As the foremost chronicler of the Stockholm archipelago, he would have been familiar with the mine on Resarö, and as a failed chemistry student he would have known about yttrium. He bluntly refused room number 13, installed himself in another chamber and set out to make gold. He was the celebrated novelist and play-wright August Strindberg, aged 47, the atomic number of silver. He was also an accomplished painter, and thus a man of many talents, self-esteem being not the least of these.*

He set up his laboratory here, not far from Marie and Pierre Curie's Sorbonne buildings, for he was going to show them all— the dusty old professors at Uppsala University, and especially that fraud Mendeleev with his proposed Periodic Table. He was going to make gold, revolutionize chemistry, and write the ultimate book about the universe—but perhaps it would have been better if he hadn't got to the end of the number series, and instead taken some lithium.

Lithium carbonate (Li_2CO_3) and other lithium salts are today a standard treatment for the manic phase of bipolar disorder. An amateur should not pronounce psychiatric diagnosis, especially

---

* To the Swedes he is the 'world famous author'; to the rest of the world he is the misogynous playwright known mostly for one drama, *Miss Julie*. This piece, however, is continuously performed around the world, and Strindberg's work with theatre and drama is considered very influential. His first novel *The Red Room* was published in 1879.

not on long dead people, but the experts seem to be in agreement that Strindberg was suffering from psychosis during this time.[96] Would he have been help by lithium? We don't know—his psychiatric diagnosis may have been different, but many people with bipolar disorder, or manic depression as it was called, declare that lithium medication helps them live normal lives. But this is not clear-cut—nothing ever is—and if anything is complicated it is the chemistry of the brain. British actor Stephen Fry, diagnosed with bipolar disorder in early middle age, admits being ambiguous about taking the drug as there are possible side effects, but also, he believes, the manic phases have actually helped him in his career.[97]

Whatever choice Fry made, it is clear that this is difficult, and every case needs to be individually judged. To a chemist it is obvious that the chemical processes in the brain can be altered or even fine-tuned by medication, or perhaps the right kind of food, but also that finding out how to do this properly is very tricky indeed. But emotional responses evoked by words and actions are also chemical reactions, and mental recovery it isn't just a matter of DNA, medication, or diet, but also of life events and circumstances.

The biochemical activity of lithium is still under investigation, but it is clear that there might be many targets. For example, Li^+ may work on the ion channels, replacing the sodium ions and slowing the signalling system down.[98] This may be related to its size. It is very small—in fact, it is the smallest metal ion with a +1 charge. The lithium ion has a radius that is 35 per cent less than for the sodium ion (Na^+)—a rather significant difference.

To me, this is one of the charms of chemistry: it can be mathematically complex, but also as simple as a child's tinker-toy set, relying on simple things like differences in size. From time to time, we take our own tinker-toy sets out of the drawer, but more often than not these days we use a computer. And just like a little child that might be fascinated by the shape, colour, and texture of a set of balls, a chemist needs to poke and touch atoms and molecules to find out what their properties are.

Now, it is a bit difficult to find something small enough to poke an atom with, but what we really want to know is how it reacts if we put a tiny charge of either plus or minus close by, and this is something we can do. A neutral lithium atom is neither soft or hard when you approach with a charge, but once you remove one electron it becomes Li^+, and just like a snail it withdraws into its shell, becoming small and 'hard'.

In what may seem to be a paradox, in the body the difference in size between Na^+ and Li^+ may actually make the Li^+ ion larger in effect, because its smallness and hardness make Li^+ hold on to the surrounding water molecules much more firmly. This occurs because the negatively charged oxygen end of a water molecule comes much closer to the lithium nucleus, and thus to the positive charge, compared to a sodium ion. The electrostatic bond (attraction of different charges) that holds the water molecules in place around the metal ion is therefore going to be stronger.* The longer the distance the smaller the interaction energy, and the bond will be easier to break. So while Na^+ ions may easily let their *waters of hydration* escape, appearing almost naked, the lithium ions will wear a heavy overcoat of water molecules at all times.

---

* There is a nice formula showing this. The energy of interaction between a dipole (a molecule having positive and negative ends) and a charge is proportional to (charge) × (dipole 'strength')/(square of the distance) or as we would write: $E = k \cdot q \cdot \alpha / r^2$. The longer the distance the weaker the energy and the bond will easier to break.

Gold on the other hand is very soft as an atom, which somehow, but not directly, has a link to the well-known physical property of gold metal being soft and malleable. You may think that it too, just like a bigger snail, should withdraw and become tough and unyielding as we remove one electron to make gold in oxidation state +1, Au^+, but no. Of course, it is a little less easy to disturb than a gold atom—after all, the overall charge of +1 means the electrons are held in tighter reins. But as the gold atom is large, the reins are long and the positive nucleus cannot easily control the outermost electrons. Because of this, Au^+ ions are easy to deform, and quite different from the lithium ions that are also much smaller, having only about a tenth of the volume of a gold (I) ion. The technical term is that they are highly polarizable.

The point of all this is that the hard guys like to play with other hard guys, and the softies will keep to themselves. By simply looking at the Periodic Table we can work out who is who, although it needs a few more tools than I have given you here. For example, we can understand why cyanide ions (CN^-) are used in gold refinement (they are soft), and perhaps how to make better lithium ion batteries.

Gold is known to poison the mind—and not by ingestion; just thinking too much about it may be enough. Actually, eating small quantities is harmless. It is a UK-approved food colouring most often found in small, delicate and rather expensive chocolate pralines. Gold compounds have also been widely used against arthritis, but have now been replaced by more efficient drugs, and this was not one of August's problems anyway.

Strindberg's mind was not poisoned by gold in the usual sense, although at this time, after two divorces and with an impressive (or rather detrimental) drinking habit, he was, in spite of his fame, living on the verge of poverty. No, Strindberg wanted to bring about his own scientific revolution, and was a rising star of the Parisian scene of the occult and esoteric.[99] Latter-day writers have made much of the fact that only 20 years later classical physics would be shattered by the advent of quantum mechanics, and in the 1940s gold would indeed be synthesized from other elements, ostensibly proving Strindberg's iconoclastic qualities also in the sciences.

These thoughts had, however, been common in alchemy for a long time and not unique to Strindberg. The idea that the elements were far too many—in 1896 they already numbered 65—and that something simpler and unifying must be lurking behind this disordered diversity was very powerful, and had been so since antiquity. As we now know, this was correct, but in the 1890s all the experimental evidence was to the contrary.

Having said this, it should be noted that within his own system Strindberg was in fact more or less rational, but very biased. Alternative explanations not conforming to his ideas were never considered. Strindberg did mostly arrive at his conclusions by experimental observations, but these were selectively fitted to more or less arbitrary combinations of numbers from the Periodic Table. He was also guided, perhaps in his less clear moments, by perceived messages from 'the Powers' in everything from writings on the wall to stumbling on the Hotel Orfila and the statue of the Spanish-born Parisian chemist and physician Mathieu Orfila, a pioneer of legal medicine, on the same day.

By another coincidence, the discoverer of lithium was also named August, more specifically Johan August Arfwedson, and at the time of his crucial analysis in 1817 he was only 25 years old. Arfwedson, in contrast to Strindberg, had not one but two degrees from Uppsala University. The curious thing is that he made the discovery in a mineral sample from Utö, another island in the archipelago of the Swedish capital, very close to Kymmendö where August the author used to spend his summer holidays. That was until he wrote a novel about the island and its inhabitants so marginally disguised that he was no longer welcome as a summer guest.[100]

It is easy to make fun of the ideas, actions, and conclusions of researchers active 100 years ago, be they mainstream or on the fringe of proper science. It is, however, much more difficult to put yourself completely in the mindset of a particular epoch and see reality as these men and women saw it. I hope I am not doing the former, and I do not have the ambition or knowledge to do the latter. I will just note that the great cultural divide (as suggested by C. P. Snow in 1959)[101] had not opened yet, and a radical rationalistic and materialistic attitude towards science was not yet fully developed as a role model for scientists. And even though they were not significant for chemistry, Strindberg's scientific experiments were important for his literary development. The future Nobel Prize winner in Chemistry, The Svedberg, wrote in 1918 that: 'his scientific studies have undoubtedly fertilized his writing to a high degree, enriched it with new and fresh images, and in an unusual way have brought him into intimate contact with the world surrounding him'.[102]

Strindberg was also apparently accomplished in practical matters. Chemist and chemistry historian George B. Kauffman tried to reproduce Strindberg's experiments in the 1980s, to the extent that he even stayed for a long time in August's last flat, the 'Blue Tower' in central Stockholm. Following Strindberg's instructions in such publications as *L'hyperchimie* he made the iron oxides and hydroxides that were obviously the end product of the 'gold synthesis'. He concludes that the synthesized material could only have persuaded 'someone as imaginative and already convinced' as Strindberg. But then he had a look at the real stuff—Strindberg's actual samples deposited at the Royal Library in Stockholm and at the Lund University Library—and had to concede: 'The samples looked much more like gold than those that I had prepared'.[102]

That they were iron oxides and hydroxides was also established early on, as Strindberg handed in his samples for confirmation (surely just a matter of form for him) by independent chemists. The results upset him so much that in one of the leading Swedish daily newspapers he publicly accused the engineer Johan Landin of having made an erroneous analysis.

Hotel Orfila no longer exists, but a plaque on the building commemorates Strindberg's six months at the hotel and the novel *Inferno* that ultimately resulted from his stay. In this book he acknowledges Orfila as his Master, referring to his chemistry book from 1817 put in his path by 'the Powers' that portentous spring and summer of 1896.

As for the dusty old professors in Uppsala, Strindberg was not completely on the wrong track. His contemporary, Svante Arrhenius, barely managed to get his PhD at Sweden's oldest university—the old professors deciding that he was clearly not cut out for such an illustrious institution—and had eventually to move down the academic ladder to the newly formed Stockholm University College. There he went on to amaze the worldwide chemistry community by showing, among other things, that in water solutions of compounds like lithium carbonate, the Li^+ ions are actually swimming around on their own (or rather, as we have seen, with an overcoat of water molecules), not as single molecules of Li_2CO_3. So, in a way, August got his revenge on the old professors by proxy when Arrhenius was awarded the Nobel Prize in Chemistry, 1903.

Was August really the last alchemist in Paris? Probably not. As a part of the esoteric and occult, alchemy is still thriving in a subculture of its own, although one suspects that, as with amateur chemists and illegal drug and bomb makers, obtaining the necessary starting materials is much more difficult today than 100 years ago. But he was probably the last famous alchemist—if we do not believe J. K. Rowling of course, who has Nicolas Flamel dying on the last pages of the first Harry Potter book in 1997. Flamel, who owned the house still standing on 51, rue de Montmorency, a stone's throw from the modern Centre Georges Pompidou in the very heart of Paris, has acquired a posthumous reputation for making the philosopher's stone, but he doesn't beat Strindberg: Flamel died in 1418.

On a final supernatural note, I should add that August's ghost can be seen and heard at the Hotel Chevillon in Grez-sur-Loing, south-east of Paris, or so I am told—even though it was in his Stockholm residence, the Blue Tower, that he passed away at the age of 63 in 1912.

In the next chapter we will turn to mental problems that we can, quite unambiguously, cure with very simple chemistry. For that we move to the element that will give you the atomic weight of gold, 197, if you combine it with the element named after France, gallium.

13

Pardon My French:
Captain Haddock and the Sufferings
of the Savoyards

In this chapter we climb to the remote valleys of the European Alps, learn what is in scant supply there, of the benefits of trade, and of a chemical bond your chemistry teacher probably did not tell you about, but that makes our minds soar above sea level.

Spending time in the European Alps means, especially in the tourist season, being constantly reminded of both the heroism and suffering brought about by this magnificent landscape. The bookshops of Grenoble all have prominent displays of the exploits and adventures of living and prematurely killed Alpinists, and on the radio you can but wait for the news of this year's first deaths on the slopes of Mont Blanc—at 15,782 feet (4,810 metres) the highest mountain peak in Europe.

But this landscape used to be cruel in a more sinister and hidden way, invisible to the naked eye. Not before certain experiments were made on seaweeds gathered on the Normandy beaches could we begin to understand and deal with the cause of the terrible sights and encounters Swiss Alpinist pioneer Horace-Bénédict de Saussure had in a small remote village near Aosta in the Piedmont in present-day Italy.

Saussure, a young professor at the university of Geneva, was out on one of his numerous hikes in the western Alps, nowadays part of Switzerland, Italy, and France, but then largely under the jurisdiction of the Kingdom of Sardinia. On this summer's day in 1768 he came upon a small village and naturally wanted to know where he was, so he asked the first man he met on the way into the village, but got no reply.

With one person, that could have been a language problem, or a general distrust towards suspicious strangers (entering a small café in a remote village and registering everyone inside going completely silent does not mean they have all simultaneously developed a speech disorder). However, as he got further into the village and still got no more than inarticulate grunts from the second and third person as well, he began to wonder what was going on. Closer to the village centre he saw a disquieting number of men and women with enormous goitres, fat lips, perpetually half-open mouths, and blank expressions, and was terrified. As he recalls in the second volume of the first serious description of this region, *Voyage dans les Alpes*, 'It was as if an evil spirit had transformed every inhabitant into a dumb animal, leaving only the human form to show that they had once been men'. He left sad and frightened, with an image he would never forget etched on his retina.[103]

Although you could meet people having the same symptoms throughout Europe, with slightly higher populations in certain, usually relatively isolated, regions (goitre was also known as 'Derbyshire neck'), you would encounter nothing like what Saussure saw any-where else. But in the Alps it was not uncommon. The extent of the problem was finally surveyed and recorded in a report to the King of Sardinia in 1848, and in Switzerland it was established that some regions had populations with up to 90 per cent affected with these physical problems, and 2 per cent showing severe mental retardation.

FIGURE 26 A Savoyard with a goitre problem, perhaps 'a crétin des Alpes' by Dominique Vivant © The Trustees of the British Museum.

The people Saussure and other visitors met—tourists in the region becoming more and more frequent by the nineteenth century*—were none other than the 'crétins' made famous in Belgian author and illustrator Georges Remi's (Hérge) Tintin books. Converting the elaborate oaths of Tintin's companion Captain Haddock to the more sensitive Swedish and English languages of the 1950s and 1960s was apparently quite a problem for translators, so 'Bougre d'extrait de crétin des Alpes' may have ended up as something quite different, but 'crétin des Alpes' or 'crétin de Savoie' is exactly what these people were known as at the time— 'the idiots from the Savoy'.

The unravelling of these two conditions, the severe mental disability that came to be known as 'cretinism', although this term is no longer acceptable, and the tell-tale visual signs of goitre, was a quest that had engaged the medical profession for as long as the conditions had been known, which seems to be from around 3,000 BC. A 'crétin' was described in Diderot's Encyclopedia as an 'imbecile who is deaf, dumb with a goitre hanging down to the waist' often found among the populations in present day Switzerland, southern France, and northern Italy.[104]

The reasons were mysterious. Parents both of whom had goitres would produce offspring with the same symptoms, so there was clearly a hereditary factor, but as for the rest, speculation ran wild. Perhaps it was the prevailing wind directions in these remote valleys? The 'Föhn' wind will make even the people of Austria temporarily insane, this is well known, so who knows what effect an evil squall

---

* Mary and William Wordsworth (mentioned in passing in Chapter 12) toured the region in 1820, and Mary writes in her diary: 'The Valais certainly must be a direful residence;—stagnant & unwholesome water, & Goitres Cretins & deformed Persons very frequent.—The very Children give you pain to look at.'

could have in these remote places? The absence of winds in some valleys producing stale air was put forward as an alternative theory, and the permanent lack of sunlight on some mountainsides during the dark months of the year was also invoked as an explanation.

Traditionally, goitre had been treated with sea-sponge ashes, and in 1811 Bernard Courtois, a Parisian salpêtrière, discovered the new element iodine in seaweed ashes used in his potassium nitrate factory. This made a physician from Geneva, the Edinburgh-educated Jean-François Coindet make a brave leap of faith, or perhaps we should call it an educated guess. In 1819 he suggested that sea sponges also contain iodine, and that intake of iodine can cure goitre.[105]

With atomic number 53, and placed under bromine in the yet-to-be drawn up Periodic Table, iodine fascinated chemists. The crystals were a deep purple colour with a distinct metallic lustre, yet it was not a metal. The air inside a glass jar of the stuff would fast develop a faint purple colour, and some of the crystals would transfer from the bottom of the jar to the underside of the lid seemingly without effort—a process known as sublimation.

Many things were still unknown about iodine and its chemistry in 1819—for example, its relationship to the upstairs neighbours bromine (to be discovered in 1826) and chlorine (discovered in 1774), but still Coindet tried it out on his patients. With too much success as it turned out, because not only did his goitre patients get better, fellow physicians adopted the remedy for completely different conditions, and soon the health-conscious citizens of Geneva were overdosing to the extent that iodine poisoning became a problem.[106]

Why would iodine work? At that time, the idea of trace elements in the body—very low concentrations of different elements (for example copper, selenium, manganese) needed for the proper function of enzymes and other molecular systems—was probably alien, and the analysis of nutrients in foodstuffs not possible as the identity of the large part of these were unknown. Nevertheless, in the middle of the nineteenth century, the Parisian botanist and physician Gaspard Adolphe Chatin analysed the iodine content of plants, soil, and water collected during his holidays in different parts of France and Europe.[107] Estimating the daily intake of various foodstuffs, he then calculated that the average Parisian stuffed himself with 5–10 micrograms of iodine per day, that is 5–10 million parts of a gram, whereas the citizens of Lyon and Torino only got 1–2 micrograms with their food and the peasants in the Alpine valleys would have to make do with less than 0.5 micrograms.[108]

He duly put forward a theory on the importance of iodine in avoiding goitre and cretinism, but with little effect. Chatin had to wait until 1896, five years before his death, before German chemist Eugen Baumann could finally show that the thyroid gland—the gland that sits on the lower part of the neck and that swells up dramatically in goitre sufferers—contains iodine. This opened up the possibility for cure and prevention on a large scale, and even more so when the molecular secrets of thyroid glands were unravelled in the form of the two chemicals simply known as T_3 and T_4, triiodothyronine and thyroxine, with three and four attached iodine atoms respectively in their otherwise organic structures.

These are very odd molecules indeed, which can be better seen if we go from the formal stylized line drawing in Figure 27, and instead try to imagine the size of the molecule and the atoms it contains. As we add protons and neutrons to the atomic nuclei to move from the lighter elements with low atomic numbers to the heavier atoms, the nuclei of course grow in size, but only marginally so, as most of

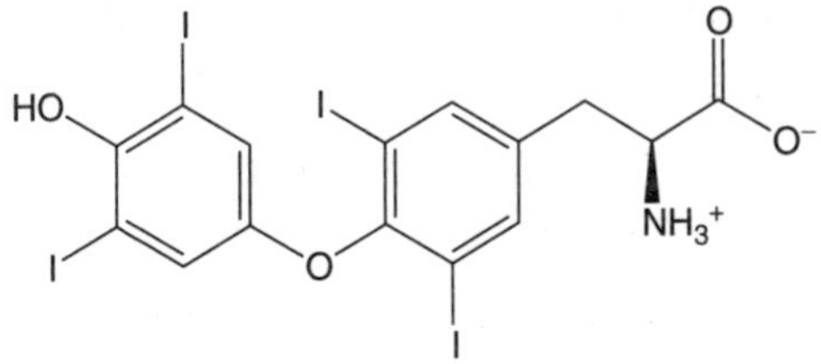

FIGURE 27 The thyroid hormones triiodothyronine and thyroxine, also know as T_3 and T_4 because of the number of iodine atoms attached to the organic skeleton. The 'fat' black bond to the NH_3^+ group signifies that this bond is pointing out of the plane towards you.

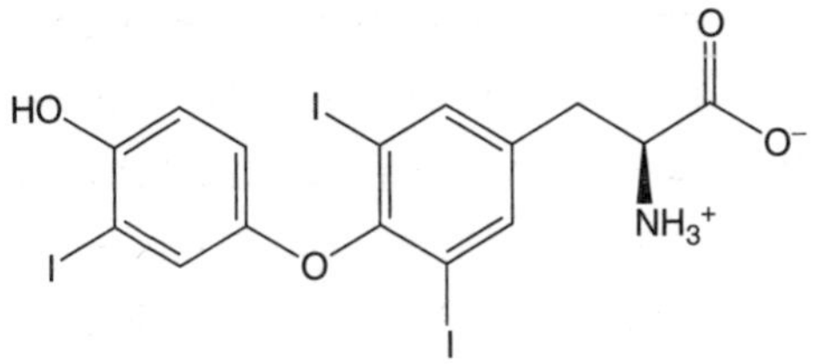

the space in an atom is occupied by electrons whooshing around. As the number of protons increases, the charge of the nucleus grows, and so the force pulling the electrons towards the nucleus also increases. This means that the atoms have a tendency to get smaller as the atomic number increases. On the other hand, the number of electrons must also increase, and as these negatively charged elementary particles are drawn closer to the nucleus they will start to repel each other, making an atom appear larger in size.

Which force will win? This is easy to figure out if you look at the Periodic Table: moving along the rows left to right, the attraction wins and the atoms get a little bit smaller; moving down a column the repulsion wins, as following each noble gas (the column of elements on the very right) the electrons tend to be added to orbitals or shells outside the electron core of the noble gas atom, and will not feel much of the added nuclear charge.

Having said this, it would not be chemistry if there were no contradictions to this rule, but that is for another chapter. It is enough to realize that having moved down the halogen ladder from fluorine (F) to chlorine (Cl) and bromine (Br), when we get to iodine we do have an atom with a size quite different from that of a carbon atom. Look at the representation in Figure 28 and you'll see how the iodine atoms, which are just like all others in the line drawing, now dominate the picture.

But size is not all. The iodine substituents as they are called—because we see them as having replaced a hydrogen atom on the hexagonal C_6 carbon ring—are in fact not nice and round with electrons evenly flowing around the nucleus; they look more like stuffed olives with the carbon–iodine bond inserted right into the reddish stuffing. The olive-green surface has attracted a surplus of electrons and is negatively charged, while the stuffing region

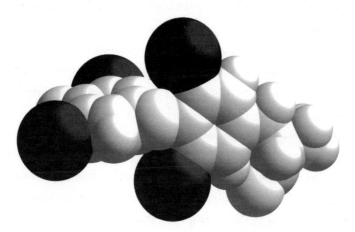

FIGURE 28 Here we have tried to represent the true size of the atoms in the thyroid hormone thyroxine, T_4, by drawing them as spheres with different sizes. Note how the big dark iodine atoms dominate the picture although they make up less than 9 per cent of the total number of atoms in the molecule, (all other atoms are in light grey).

has been deprived of electrons and the positive nucleus shines through, giving this part a positive charge.

This means that an atom, usually an oxygen, having a pair of electrons sticking out in a very pointy way—and they tend to do this in the amino acids forming our proteins—can pick up the T_3 and T_4 molecules by pointing this electron stick into the red positive stuffing forming a weak chemical bond. This is how the thyroid hormones fulfil their important missions in the body: by passing a message around, not only regulating neural and sexual development and growth, but also controlling the chemical routes by which foodstuffs are broken down, energy extracted, and heat produced.

Obviously a correctly functioning thyroidal gland is of utmost importance to our well-being, and you can have all sorts of thyroid-related problems and symptoms. However, let us for the moment stick to what was, in the early part of the twentieth century, correctly identified as a deficiency rather than an illness—goitre and 'cretinism'.

In Switzerland all salts for human consumption have been iodized since 1922, but other governments were more cautious. The French Academy of Sciences had warned against overdosing on iodine in the 1860s, and only in 1952 was table salt with added potassium iodide (KI) allowed in France. The problems in the Alpine *départements* nevertheless seemed to disappear by themselves, the invisible hand of market economy distributing iodine in the forms of imported groceries from regions with better soils, notably fish and seafood products that are normally very high in iodine.

As we now easily access villages and former *alpages* in the Alps to go skiing and hiking, it is easy to forget that even in the nineteenth century, very close to old and large urban centres like Grenoble, there were populated areas that could be reached only by narrow donkey paths over forbidding mountain passes, and this only in summer, while the rest of the country was rapidly becoming connected by modern railroads. Civil engineering projects that did much to solve sanitary conditions in big cities, by letting debris and wastewater flow out in modern underground tunnels, also enabled iodine to flow into remote mountain villages through roads blasted through the rock with new and relatively safe explosives.

Why then is this precious resource so unevenly distributed on earth? Rains and the constant outflow of water in the Alps have slowly eroded the soil of its iodine content as the simple sodium and potassium salts (NaI and KI) are very soluble in water. The same may happen in coastal regions with a lot of rainfall, such as Bangladesh, but generally speaking the mountain regions of the world are the worst hit.

And I say 'are' because, unbelievably, this is still a problem, despite all we know and how simple the solution seems to be. It has been estimated that between 800 million and 2,200 million people worldwide are still affected or at risk of iodine deficiency today, and a few years ago the *New York Times* ran a headline saying: 'In Raising the World's I.Q., the Secret's in the Salt'. The article tells of the struggle and success of raising the amount of iodized salt consumed in the former Soviet republic of Kazakhstan, and one who helped was former Chess World Champion Anatoly Karpov, a long-time crusader against iodine deficiency in his role as UNICEF ambassador.[109]

These efforts are continuing worldwide, and are often helped by advocacy from The International Council for the Control of Iodine Deficiency Disorders (ICCIDD), an organization with official recognition from UNICEF and the WHO, and a partner in the Network for Sustained Elimination of Iodine Deficiency. It has been a partial success. In the 1980s only 20 per cent of the world's population could cook with iodized salt; today, the figure is around 70 per cent.[110]

But we have a fair distance to go, and iodizing salt is the cheapest and easiest way to combat this type of brain damage in children. Still, it is estimated that 18 million babies are born with a mental impairment every year because of their mother's inadequate intake of iodine and iron, another global micronutrient problem. Even in Western Europe as many as 50 million people may be on the borderline of slight deficiency.

As for the actual iodine needs, it should be noted that Gaspard Chatin probably underestimated the daily dose of iodine of the 1860s Parisians, or they too had grave iodine problems. The recommended daily intake today is 100–220 micrograms per day, and even more for pregnant or breastfeeding women, compared to Chatin's estimated 10 micrograms per day.

Finally, while Horace de Saussure was one of the first to widely publicize the cretinism and goitre problem in the Alps, he was also the first to promote the climbing of Mount Blanc, announcing an award for the first to reach the summit in 1760. He himself was the third man on the peak in 1786, a year after Jacques Balmat and Michel-Gabriel Paccard. The first woman to conquer the mountain seems to be a somewhat disputed story, with the contestants being Maria Paradis and Henriette d'Angeville, reaching the summit in 1808 and 1838 respectively.

14

Two Brilliant Careers

*In Chapter 14 we explore a connection between murder and moun-
tain, in a story in which the chemical concept of solubility is used to
solve a crime, and to make efficient pills for the pharmaceutical
industry.*

At the time of publishing, it is exactly 50 years since Bob Dylan
answered a number of enigmatic questions with the ambiguous line 'the answer is blowin' in the wind' on the A-side of the
record *The Freewheelin' Bob Dylan*. But one of these, 'How many
years can a mountain exist before it's washed to the sea?' we can at
least try to answer, as part of the solution lies in one of the more
famous rules of thumb one learns as a novice chemist: positively
charged metal ions combined with oxides (O^{2-}), sulphides (S^{2-}),
phosphates (PO_4^{3-}), silicates (SiO_4^{2-}), and carbonates (CO_3^{2-}), are
insoluble in water, whereas similar nitrates (NO_3^-), chlorides (Cl^-),
and bromides (Br^-), are soluble.*

---

* Normally the complete list goes like this: oxides, hydroxides, sulphites, sulphides, phosphates, and carbonates are insoluble; nitrates, acetates, and sulphates
are soluble. In the real world there are of course numerous exceptions to these
rules.

In terms of stuff you've got in your kitchen, this means that when you put a spoon of table salt (NaCl) into water it will 'disappear', faster if you stir or heat, and the water will look exactly the same as before. For the insoluble stuff, we move to the more expensive regions of the cupboards and investigate the state of our silver and copperware. When things like these were to be on display, my mother used to have me clean them with silver or copper polish, as the oxides and sulphides tarnishing the metal surfaces did not go away in a normal wash with water—they are completely insoluble. A suitable but boring exercise, and as close to a chemistry set as I ever came as a child.

What they normally don't tell you in chemistry textbooks however, are the enormous consequences of these rules, visible all over the world. Why are mountains made of rocks from oxides, sulphides, phosphates, silicates, and carbonates? Because they are insoluble! Any mountains made from sodium chloride would indeed have been 'washed to the sea' thousands of years ago, and where NaCl can be mined it is also known as *rock salt*, and found either underground or in regions with a very dry climate.*

Which brings us to the hero and heroine if this chapter. As far as I know, Agatha Christie (1890–1976) and Herbert Dow (1866–1930) never met, but they both partly owe their brilliant careers to the same element, bromine, with the Periodic Table address atomic number 35, group 17, period 4, symbol Br, the immediate downstairs neighbour of chlorine.

---

* Large deposits are also found, somewhat unexpectedly, underwater, for example at the bottom of the Mediterranean Sea—evidence that these seas and lakes were once dry. Why don't they dissolve? A saturated salt solution is very dense, and will accumulate on the bottom and mix very little with the bulk water of the ocean, thus protecting the salt layer from being dissolved.

As the bromide ion (Br^-) is the only form (or rather oxidation state) in which this element occurs in nature, the solubility rules tell us that mining for bromide minerals is unlikely to be successful. Compounds like $NaBr$ are very water soluble, and bromine is also a heavy element and these are, as a very general rule with many exceptions, less common than the lighter ones. However, the closeness in properties between elements in the same group (column) means that wherever we find chlorine we may also expect to find bromine, but in smaller concentrations. Indeed, in seawater we find 666 chloride ions for every bromide ion (this number looks sinister, but I am confident this is just a coincidence, not a message from a higher authority).

Herbert Dow, a chemist and inventor with a shrewd sense for business, devised his own clever way to extract bromine from *brine* (concentrated saltwater) pumped up from a huge underground lake in Midland, Michigan, in the US—a project he had already started as a student at the institution today know as Case Western Reserve University. At that time, in 1891, bromine production was a recent development. The relatively newly discovered element had only been available in large quantities for about 30 years. Bromide was now in demand for both the newly established photographic business, as it is one of the starting materials to make the light-sensitive compound silver bromide ($AgBr$), and, more surprisingly perhaps, for the rapidly expanding market for 'patent medicines'.

In a world in which new things were good things, as long as they did not interfere with the established social order, a buccaneering attitude towards newly prepared compounds prevailed. Applications could be found and quickly marketed. Queen Victoria's obstetrician, Sir Charles Locock, suggested the bromide ion as a treatment for epilepsy, and although inefficient by today's standards, potassium bromide (KBr) was the first means physicians had to control this chronic disease.

It was soon realized that the bromide ion had sedative effects in general, and it was used by trend-setting physicians such as Jean-Martin Charcot in the Salpêtrière Hospital, in Paris (an immense, still existing complex next to the Gare d'Austerlitz). One of his patients was a Parisian delivery man who had bizarre, reoccurring, and sometimes long-lasting spells of lack of awareness that still let him interact with people and ramble around town, and even to take a train as far away as the coastal town of Brest. From Charcot's case notes, it seems he was partly cured by repeated bromide therapy, having long symptom-free periods and only shorter relapses while on medication. However, bromide seems to have been more symptom-relieving than curing, as less than a year after his final treatment the man wandered off, never to be heard of again.[111] This partial success of Dr Charcot, and other such stories, combined with the severe lack of suitable therapeutic tools for physicians in the 1890s, led to an almost hysterical over-prescription of KBr and NaBr by doctors. Some specialized hospitals apparently used tons of the stuff every year,[112] which meant a thriving business for the pharmaceutical industry (then in its very early infancy) such as that of Dr Miles. (Charcot's patient was on 7 grams a day at his highest dose; we are not talking about the 0.2 g of active ingredient of an ibuprofen pill.)

This is not likely to have disturbed the sleep of Herbert Dow who, presumably, did not need to use Dr Miles' Nervin preparations, made in Elkhart in nearby Indiana, as the Dow Chemical Company was doing very well indeed.[113] A fair guess is that the bromide salts used by Dr Franklin Miles—the first product of

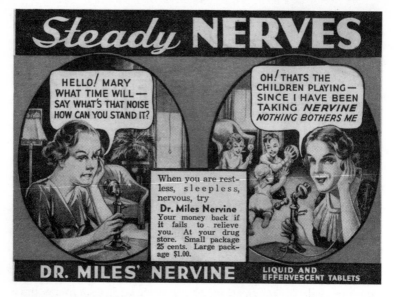

FIGURE 29 Advertisement for Dr Miles' Nervine, a preparation based on bromide salts such as NaBr. From *Dr. Miles New Jokebook*, 1933.

what was to become Miles Laboratories, a renowned US pharmaceutical company—were bought from Mr Dow's factories.

A more remarkable business accomplishment by Dow was his confrontation with the mighty Deutsche Bromkonvention, a cartel of German companies with a near monopoly on the international bromine trade. The Germans flooded the US with cheap bromine in order to undercut Dow when he tried to venture into the UK and Japan. Dow, however, had enough financial muscle to buy up the German bromine clandestinely in the US and re-export it to Europe, including Germany, eventually turning the tide on the mighty cartel.[113]

Someone who did apparently suffer sleeping problems however was Mrs Inglethorp, the mistress and tyrant of Styles Court, a remarried widow with a number of potential relatives looming around the estate, each with more or less convincing claims to a coming inheritance. This insomnia is used as an excuse for former pharmacy student Agatha Christie to put a solution of Dr Miles' Nervine, or some other similar bromide concoction, on the dressing table in Mrs Inglethorp's bedroom, conveniently at hand for her murderer.

If you have not read *The Mysterious Affair at Styles*, Agatha Christie's first novel, published in 1920,[114] you can stay calm and read on. I will not give away the plot—only tell you how the murder was committed. Perhaps you are cleverer than me, and this information will help you to identify the murderer before Hercule Poirot, her now famous Belgian detective, in classic manner invites the main characters to a 'little *réunion* in the *salon*'. In that case I beg you to excuse me, but I will gladly confess I was still very uncertain at that point, as the novel contains enough red herrings to feed an entire CID team.

Many of these smelly fish concern *strychnine*, a substance everybody in the novel seems to have had access to. As it turns out, the fatal dose came from a bottle in Mrs Inglethorp's own bedroom, as in addition to the calming bromide ions she also had a prescription for strychnine: in very low doses a stimulant to the central nervous system.*

A small sip from that flask every night would be completely harmless, but if the entire contents were to be consumed at once it would be fatal. The murderer comes up with an ingenious plan to make the very rich widow do just this—take the entire lethal dose by her own will.

---

* No longer in use, but as a modern book on pharmacology says, the thinking was probably: 'it tastes so bad that it should obviously do the patient some good'. *Pharmacology for Health Professionals*, Bronwen Bryant and Kathleen Knights, Mosby, 2006.

The plan, of course, has to do with solubility, and as it turns out, solubility is a common problem for drug molecules—pharmacologically active compounds. Even though pharmacists do not use strychnine today, it is a good example. Like many extremely toxic substances, strychnine is naturally occurring, and it can be extracted from the seeds of the strychnine tree (*Strychnos nux-vomica*) native to India.[115] In the extracted form, it is however useless for preparing 'tonics', as this fat-loving molecule is completely insoluble in water—a characteristic it thus shares with many active drug components.

Look at Figure 30, and you will see that carbon–hydrogen and carbon–carbon bonds dominate the strychnine structure. These bonds are all made from a fairly even sharing of electron pairs between the atoms—very different from the situation in a water molecule, where a greedy oxygen atom keeps the two electrons it formally shares with each hydrogen close to itself, making these atoms distinctively negative and giving the hydrogen atoms a positive charge. We call water a polar molecule because it has two ends with high and distinctively positioned charges, and strychnine

FIGURE 30 Left: the strychnine molecule, insoluble in water. Middle: the strychnine molecule when it has acted as a base and taken a proton (H⁺, black) from hydrochloric acid (HCl) forming the salt (strychnineH⁺)(Cl⁻), known as strychnine hydrochloride, that is water soluble.[116] Right: line drawing of strychnine.

a non-polar molecule because no such substantially charged centres exist in this structure.

Another basic rule of thumb an emerging chemist learns is that 'like dissolves like', something he or she may already have picked up in the kitchen. Table salt has a very polar structure, and dissolves into sodium cations (Na^+) and negatively charged chloride anions (Cl^-), which both love to splash around and play with the polar water molecules. On the other hand, many flavour molecules are also fat-loving and dissolve badly in water. This is why a bit of cream or oil enhances the taste of many dishes—these wonderful molecules are dissolved in the fat, and can then be easily transported to our olfactory organs. Moreover, if you are the dieting type, you should know that the same goes for some vitamins, notably vitamin A—they also dissolve in fat, so carrots with butter or olive oil may not be such a bad idea.

On a molecular level, we roughly divide our substances into polar and non-polar, and non-polar compounds are not soluble in polar solvents like water. This is much like Velcro or felt compared to a flat glass surface. Felt sticks to felt but not to glass, and two flat and clean glass surfaces stick to each other but not to Velcro, the flat and even surface being the analogue of the non-polar C–H and C–C bonds, and the hooks and loops of the Velcro corresponding to the negatively and positively charged ends of a polar molecule.

Some drug molecules are so non-polar and insoluble that if you put them in a pill and have the patient swallow it, even the vicious acidic conditions of the stomach will not be sufficient to release them into the blood stream before they are excreted out, making such a tablet practically useless.

A lot of effort is therefore put into circumventing this problem by the pharmaceutical industry, as these companies do love pills, and one classic and successful strategy is to make a non-polar molecule polar by adding a charge to it. The way to do this with strychnine is to add one molecule of hydrochloric acid (HCl) to every molecule of the potential poison. This works because at the 'top' of the molecule, as I have drawn it in Figure 30, there is a black nitrogen atom sticking up. This part of the molecule is a *base* (as opposed to an acid), more specifically an amine related to the simple chemical ammonia (NH_3) which you may be familiar with from the cleaning cupboard. It has a bit of a negative end sticking up from the non-polar bulk of the molecule because of a pair of electrons that are positioned there, pointing straight out from the nitrogen atom. This bit of 'polarity' is not enough to make strychnine water soluble, but add HCl and an H^+ will immediately jump to this position, and suddenly the strychnine has a charge of +1 with a chloride ion to balance the charge, and now the molecule is ready to go swimming (but will retain its chemical activity). Similar reactions make the drugs codeine, ciprofloxacin, and many others more water soluble. On the ingredient list this will be stated as, for example, 'codeine hydrochloride salt', as this is indeed a reaction between an acid and a base giving a salt, just like the reaction HCl + NaOH giving NaCl and H_2O.

From what we have learnt so far, we would expect the hydrobromide salt of strychnine also to be water soluble: the H^+ is the same and the bromide ion is very similar to the chloride ion (for example, the arrangement of the ions in solid NaBr is exactly the same as in NaCl). But we are in for a surprise: a crystal of strychnine hydrobromide has a very different arrangement of ions from that of strychnine hydrochloride. Even though the anions and cations will like water about as much in both cases, the hydrobromide salt appears to have much stronger bonds between the molecules* than the hydrochloride, and therefore it will not dissolve much in water. Not because it is afraid of water, but because it finds it much cosier on the beach with its friends.

How then did this help the clever murderer? He (or she) figured out that Mrs Inglethorp had two flasks in her bedroom, one containing strychnine hydrochloride, and another one containing sodium bromide. Add the bromide to the strychnine solution and the protonated strychnine molecules will soon find their cosy bromide chums and settle down on the beach—or rather, as a solid precipitate at the bottom of the flask leaving the sodium and chloride swimming around in the water. As a small amount of the strychnine tonic is consumed every day, but now virtually void of any strychnine molecules, the level of the liquid will gradually decrease until the fatal moment when the last dose of the flask is emptied, now containing all the strychnine present in the original solution, a lethal dose as it turns out.

This was the start of Agatha Christie's long and brilliant career in crime fiction. She began writing while Conan Doyle was still turning out Sherlock Holmes mysteries, and wrote her last novel three years after Reginald Hill had introduced his famous Dalziel and Pascoe Yorkshire detectives. She worked in a hospital dispensary in the later part of World War I, where she kept herself occupied with her first novel when not preparing drugs or studying chemistry for a pharmacy exam. Chemistry, after some initial 'bewilderment' with 'the Periodic Table and Atomic Weight', was something she apparently rather enjoyed, and Christie is probably

---

* We call these intermolecular bonds or interactions.

one of the very few crime writers who actually performed a Marsh test for arsenic in real life.[117]

Dow Chemicals, still with corporate headquarters in Midland, Michigan (population 41,863), is now one of the two or three, depending how you count, largest chemical manufacturing companies in the world (BASF and DuPont are the other two). Miles Laboratories was bought by German pharmaceutical and chemical giant Bayer in 1979 and is no longer an independent company.[118] All bromide* preparations for tranquilizers were withdrawn in the US in 1975 because of long-term health problems related to the overconsumption of bromide ions, and more efficient remedies are now available.

---

* Bromide ions may still be present in some drugs as a counter ion to improve solubility, it is not toxic and on the approved list of such ions routinely tested when formulations of new compounds are tried out.

15

War and Vanity

In this chapter we encounter the first deliberately designed biotechnology process not involving the preservation of food, debunk a persistent creation myth, and learn how to look out for explosives.

In my childhood, visits to Gothenburg would always include a long (it seemed at the time) tram ride with my mother, from the centre of town to the north-eastern districts, past the old, red brick, ball-bearing factory of SKF to the vast Kviberg Cemetery to put flowers on my grandmother's grave. I never ventured on any longer excursions among the neat flower-decorated graves on these well-kept lawns, but had I done so I would perhaps have discovered a different, more uniform, part of the cemetery that relatives seldom visited: the war graves.

War graves form a somewhat unexpected discovery in the suburbs of a country that was neutral in both world wars, but there it is. Among the mostly German, American, and British graves we find, in the Commonwealth section, that of Arthur Cownden who, at 17, was probably the youngest to be buried there.[119] He was boy telegraphist on a Royal Navy destroyer, and on the morning of 1 June 1916 his body was washed ashore close to the small fishing village of Fiskebäckskil on the Swedish west coast. His ship, the *HMS Shark*, was one of many British losses during the preceding day's Battle of Jutland—the only clash between the main forces of the Royal Navy and the German *Hochseeflotte* during World War I.

FIGURE 31 Arthur Cownden's grave in Gothenburg. Photo by the author.

By all accounts this was a terrible battle, with loss of lives in the thousands on both sides, and one of the largest naval battles ever fought. The Battle of Jutland remains somewhat controversial for two reasons: the enduring argument between the two British commanders, David Beatty and his superior John Jellicoe, and the purported role of the Royal Navy's smokeless gunpowder *cordite* in the sinking of a number of its own ships.

We have no business with naval tactics, but the cordite question is related to one of the lesser-known supply problems of World War I, that of acetone. You may be familiar with this molecule as nail varnish remover, but perhaps you also know the disastrous effect it has on the glossy surface of cars. This suggests a molecule that is a very good solvent, and as such it is used extensively in the paint and pharmaceutical industries. These are all later applications of course, and both car paint and nail polish were only really developed after World War I.

Although highly flammable (nail varnish remover bottles should carry a safety label), this small organic molecule, the simplest of all ketones with formula CH_3COCH_3, is not in itself an explosive. You do not risk blowing up your boudoir by storing it with your other chemical vanity and hygiene products (and they are all chemical, even if it says 'organic' on the bottle).

How do we know it is not explosive? Well, as one of the staple products of the chemical industry for more than 100 years, we know that from experience, but there are also simple rules that enable a chemist to have ideas about the properties just by looking at the formula of a molecule.

In the case of explosives, there are three things to look out for, presuming we are dealing with simple compounds composed only of nitrogen, oxygen, carbon, and hydrogen. First we would check

FIGURE 32 Two representations of the acetone molecule, simplest of all ketones with formula CH_3COCH_3.

the shape of the molecule: does it look like a car suspension spring compressed to the maximum, ready to release its stored energy in one violent bounce? (Innocuous as they may appear, mechanical devices like these actually kill people.)

Next would be simply to check the percentage of nitrogen atoms in the molecule: the higher number, the higher risk of an explosion. Why is that? Because all nitrogens have in irresistible urge (the chemist would say thermodynamic driving force) to combine with other nitrogens to form N_2, dinitrogen gas, the major component of air. As you perhaps know, dinitrogen will react with nothing under normal circumstances—it is like a stone that has rolled down to the bottom of a deep valley. Finding a molecule with a high nitrogen count is like finding a big rock a fair way up a steep hill, with only a few small pebbles stopping it from tumbling down and leaving havoc in its wake.

Finally there is oxygen, but now it is not so much the number of atoms (sugar contains one oxygen per carbon but is not an explosive) as their neighbours in the molecule we need to look out for. Is an oxygen bound to another oxygen? Then watch out, because you are dealing with a peroxide, the simplest of which is hydrogen peroxide, a molecule you can buy diluted in water but nevertheless carrying a safety warning. Is the oxygen's nearest neighbour a nitrogen atom? Even worse, are two oxygens bonded to the same nitrogen? Then it may be time to take precautions, because you are dealing with a nitro compound, the most famous of which are nitroglycerine and TNT, 2,4,6-trinitrotoluene.

Now, if acetone is not an explosive, what was its value to the Navy? As a solvent of course: you may have already guessed that. The cordite mentioned earlier was not a high explosive used to cause death and destruction after the projectile had penetrated the enemy's armour, it was the propellant that made the shells (or bullets) fly in the first place. As such, rather than a stone tumbling down the hillside at ever increasing speed, it should behave more like a ball following a serpentine road down the mountain at a controlled pace.

It turned out that such a propellant could be made from a suitable mixture of nitroglycerine and nitrocellulose (cotton gunpowder—note the nitro prefix), with a small amount of Vaseline as a stabilizer. You will need these individual components to be well mixed though—local pockets of pure nitroglycerine may set the whole batch off. How to achieve that? Nitroglycerine is a liquid, Vaseline a jelly-like substance, and cotton fibres are evidently solid. One could imagine throwing them all in a blender and running this at high speed. This would normally produce a good enough mix, if it were not for the shock-sensitive nature of the nitroglycerine. Better then to find a good solvent for all components, and this is where acetone comes in. With this solvent a relatively homogeneous mixture could be obtained, and the dough-like product made into thin cords (thus the name) that let the surplus acetone evaporate.[120, 121]

So, where do all these ingredients come from? We can't grow plants that produce nitrated cellulose, and there are no rivers of acetone and nitroglycerine to tap into. But we can grow cotton and other plants to get the cellulose, glycerine is a by-product from soap making (ultimately originating from plant or animal fats), and nitrate was shipped in from Chile (Chilean saltpetre). Acetone was the problem kid. It was made from wood, but up to 100 kg were needed in order to get 1 kg of acetone. This is what a chemical engineer would call a yield of 1 per cent, and then sit down and cry. This was of course lousy, even in those days before green chemistry, atom-economy, and sustainability, and there was no way this process could satisfy the demands of the Royal Navy Cordite Factory in Holton Heath, Dorset. Added to which, the Royal Navy itself had already been busy chopping down the English forests during the past 400 years to make ships, and what was left had fuelled the Industrial Revolution, so the problem was indeed severe and acute.

What follows is a story riddled with unverifiable anecdotes and conflicting evidence given by some of the key individuals in their autobiographies, notably David Lloyd George—at that time the Minister of Munitions.[122] Was there a chance mentioning of 'acetone' during a lunch between Lloyd George and the Manchester Guardian editor C. P. Scott, neither of the gentlemen actually knowing what they were talking about? And what was the role of the Nicholson's Gin Distillery in the outskirts of London?[123] Either way, the consequence was one of the first successes of modern biotechnology, before the word was even invented.

The point is that acetone is a truly 'organic' organic molecule, turning up in the basic metabolism of almost every organism. We ourselves produce acetone when we burn stored fat to get extra energy. If we function normally, the process does not stop there, and eventually, after passing a number of enzymes in our body, the end products are carbon dioxide and water. The smell, however, is somewhat sweet and very distinctive, so that a clever physician may detect metabolic disorders by the elevated acetone content on the breath of a patient.

Whether Chaim Weizmann, chemistry lecturer at the University of Manchester, used his nose to detect the acetone produced by the different breeds of bacteria that he had for many years tried to persuade into making butanol from starch, we do not know. In any case, he was not the first to discover acetone in such a bacterial brew. What he did was to find the best bacteria for the task, *Clostridium acetobutylicum*, which could fairly quickly turn 100 kg of molasses into 12 kg of acetone.[124] This is still only a moderate yield, and nobody knew what to do with the large amounts of butanol produced simultaneously (two such molecules were produced for every acetone molecule). Overall it was a success, however, and made both Lloyd George, and the First Lord of the Admiralty, Winston Churchill, happy. So happy in fact that Weizmann soon became the scientific director of the British Admiralty Laboratories.[125]

It has been claimed, perhaps most elegantly in the play *Arthur and the Acetone* by George Bernard Shaw,[126] that the 1917 Balfour declaration, and consequently (although this is not explicitly stated in the document) the foundation of the state of Israel, was a gift to Weizmann in recognition of this critical wartime effort. Arthur Balfour was the UK foreign secretary during the end of the war, and Lloyd George also tells a story to this effect in his *War Memoirs* from 1933,[122] while Weizmann himself denies this turn of events in his 1949 autobiography *Trial and Error*.[127] Moreover, it seems clear that Weizmann had met Balfour many years before the war, when he had just moved from Switzerland to take up his position at the University of Manchester and Balfour was his local member of Parliament. He was already an avid proponent of the Zionist movement, and continued to work for this cause all his life, crowning his achievements by becoming the first president of Israel in 1948.

Today, the large-scale manufacture of acetone is mostly accomplished by reacting two products from the petroleum industry, propene and benzene, together with oxygen from the air. At the time of World War I this huge industry, based on oil and natural gas, was just in its infancy, and it is food for thought that the biotechnology industry, that we today regard as so promising, scored its first triumph almost 100 years ago. (That is, if we disregard the production of preservatives such as ethanol and various acids by traditional fermentation, the very foundation of this technology, and the story of Chapter 16.)

Eventually, we will run out of oil, so we could perhaps do worse than dust off the old methods of Weizmann, more so as the vast amount of butanol also produced is now an important and valuable product for the chemical industry.

Arthur Cownden's unlucky ship, *HMS Shark*, was under the supreme command of Rear Admiral David Beatty, who saw two of his battle cruisers sink after 20 minutes during the very first stages of combat. He harshly commented 'There seems to be something wrong with our bloody ships today'. Some people have claimed cordite was at the heart of the problem, as cordite stacks were causing further damaging explosions after German hits that the ships normally should have been able to sustain. Even though marine archaeologists have recently examined the wreck of one of

FIGURE 33 Professor Chaim Weizmann(front), chemist, biotechnology pioneer, and Zionist politician with friends in high places. Courtesy and copyright the Weizmann Institute.

Beatty's ships, the *HMS Queen Mary*,[128] it will be difficult to ever reach a definite conclusion on this matter.

It seems that the strict rules for the handling of the silk bags containing the cordite were disregarded on some ships, or that these rules were impeding the requested firing speed of the guns. As for the role of acetone, we know that the *HMS Queen Mary* had both the older *cordite Mk.I.,* and the improved version *cordite M.D.* with a much lower nitroglycerine content, in its munitions stores.[129] Whether this was because the *Mk.I.* was considered safe enough, or because the acetone shortage made it too expensive to replace it with the *M.D.* variety is not clear. Anyhow, the principal problem with *Mk.I.* seems not to have been explosion risks, but rather erosion of the gun barrels.

HMS Shark had no direct cordite problems, but was sunk while 'fighting to the last' as the *Encyclopaedia Britannica* put it in the 1960 edition. Its captain, Commander Loftus Jones, was awarded a posthumous Victoria Cross. Six survivors were picked up by a Danish ship that landed in Hull, but its (almost certainly) youngest crewmember is never mentioned.[130] This 31 May I again took the tram out to the Kviberg Cemetery, and put flowers on both my grandmother's and on poor Arthur Cownden's graves.

Finally, I'd like to take up a thread from a preceding chapter and ask: what produces the damaging force of an explosion? The answer is again in the Gas Law that we met in Chapter 3, and in the decomposition reaction of, for example, nitroglycerine. Chemists would write it like this, where '(l)' means liquid form and '(g)' means a gas:

$$4\,C_3H_5(NO_3)_3\,(l) \rightarrow 6\,N_2\,(g) + 10\,H_2O\,(g) + 12\,CO_2\,(g) + O_2\,(g)$$

Four molecules of liquid nitroglycerine will give 29 small gas molecules of different types. The volume of these molecules in the gas phase we can calculate by using the Gas Law in the form:

$$V = \frac{nRT}{P}$$

The water will also be in gaseous form, because a large amount of heat is generated at the same time—this is an *exothermic* reaction. (With some additional data we could even calculate exactly how much heat is generated using the reaction formula above.) If we take a teaspoon (5 ml) of liquid nitroglycerine we can calculate how many gas molecules will be generated (n), and we know the pressure (P), and the temperature (T).* It turns out that these 5 ml

---

* For those interested in the details we use the Gas Law in the form $V = nRT/P$ with the gas constant R = 0.082 dm³ atmospheres/(kelvin moles), T = 398 K, P = 1 atmosphere. With the density and molar mass of nitroglycerine of 1.6 g/cm³ and 227 g/mole we get 8 grams of nitroglycerine, which gives 0.035 moles. The total number of moles of gas molecules becomes 29·0.035/4 = 0.255 and the volume, V = 0.255·0.082·398/1 = 8.3 dm³.

will generate, in a fraction of a second, more than 8 litres of hot gas—a tremendous expansion in volume that would literally blow anything in its way to pieces.

The question of nail varnish remover for the war effort in Wold War I went up to the highest circles, as we have seen. Before the war, most acetone was imported into the UK from the very efficient German chemical industry, a fact of strategic importance that was seemingly overlooked. This is odd, as the raw materials for explosives production had been on the agenda for many hundreds of years. The next chapter will take you from the beauty parlour, with its distinct smells and odours, to another place your olfactory sensor (in other words your nose) could probably easily recognize: the stable.

This chapter is dedicated to the memory of my great uncle, enlisted seaman Bertil Johansson of the *HMS Astrea*, who died in the service of the Royal Swedish Navy in 1917, aged 18.

16

When State Security was a Stinking Business

This chapter takes us to the darkest of The Small Lands, the forest region that once was the natural border between the relatively civilized Denmark in continental Europe and the Kingdom of Sweden. Here we will learn what was brewing in the barn.

It is spring 1708, and Sweden has been at war for eight years. Charles XII camps out with the army in Lithuania, still a year from the fatal battle of Poltava in Ukraine, and it is a busy time for seasonal workers Per Larsson Gässaboda and Esbjörn Persson Bölsö. In the southern province of Småland (The Small Lands), the former border region with Denmark now just north of the new Swedish province of the recently occupied Skåne, the cows are out of the barns in which they have spent the cold winter, and it is time for Per and Esbjörn to take out their shovels, load their wagon, and set out on their mission for the King to the farmers of the region.

They are part of the army, enrolled men, but not for combat because they are *petermen*, or 'sjudare' (simmers) as they were called in Swedish. The farmers do not look forward to their visits as these men can command their chariots and their horses at will, take the firewood (and they need huge quantities), and wreak havoc to barns, stables, and houses in their quest for the manure and urine-rich soils that form the valuable raw material for their trade.

These men make nitrate—or to be specific, potassium nitrate (KNO_3), also known as saltpetre—for delivery to the King's gunpowder factories. More than 100 years ago Henry VIII's contemporary, the equally shrewd and ruthless King Gustav Wasa, had realized Sweden's precarious situation when it came to gunpowder, and with a simple stroke of his pen ruled that the soil underneath barns, stables, and cowsheds belonged to the King. In an additional law, perhaps more illustrating his fear of being cheated by innovative farmers than his well-known attention to detail, he also banned any building housing livestock from being paved with a stone floor. In a country in which buildings of stone were virtually unheard of, except for housing the very rich, this was hardly likely anyway, but the King didn't like to take chances with money and the saltpetre was a valuable commodity that otherwise would have had to be imported.[131, 132]

This particular spring day Per and Esbjörn were busy in the barn of the innkeeper Jöns Jönsson in Älmhult, a small village that would later see the birth of IKEA. The soil was normally dug up to a depth of about 1 foot, and then put into big basins with water for some time to dissolve the potassium nitrate. When the debris had settled down, a clear water solution could be transferred to big copper cauldrons containing about 800 litres each. Per and Esbjörn would then command the firewood to be fetched, and the cauldrons would be heated to boiling and left simmering for up to a week, 'or until an egg would float on the surface'. Ashes were then added to the hot solution producing what chemists would call a precipitate. In more profane words, solid junk would appear in the cauldrons and fall to the bottom. One of my sources tells me that the operators should have waited until the temperature reached 25°C. However, a temperature scale of any sort, not to mention a thermometer, was out of bounds to the petermen. They would have had to rely on their own good judgment of 'process parameters' such as temperature and viscosity, before the solids could be separated from the solution. As the temperature decreased further, crystals of potassium nitrate would appear.

The yield of the colourless crystals of Per and Esbjörn's labour was dependent on two processes of which there was very little understanding at the time. One was probably the first biotechnology process to be used for purposes other than preserving food. So-called nitrifying bacteria in the soil would feed on the degradation products of proteins in manure and urine, such as urea (H_2NCONH_2) coming from the nitrogen-containing peptide bond that links our amino acids into proteins.

This is reminiscent of what happens in the latter stages of cheese making, and explains the ammonia (NH_3) and sometimes downright farmyard odour from a well ripened Brie or Camembert. To get from NH_3 to NO_3^- is chemically more complicated, as it involves a complete change of dress for the nitrogen atom, from having three extra electrons (what the chemist would call 'oxidation state –III') and being surrounded by three hydrogen atoms, to become a nitrate (NO_3^-) ion lacking five electrons (thus oxidation state +V). Below we see the simple reaction when urea, the main and odourless nitrogen component of urine, becomes ammonia and carbon dioxide:

$$H_2NCONH_2 \text{ (s)} + H_2O \text{ (l)} \rightarrow CO_2 \text{ (g)} + 2 NH_3 \text{ (g)}$$

(where (l), (g), and (s) signify liquid, gas, and solid respectively).

It is easy to envisage the nitrogen atoms grabbing one hydrogen atom each from the water, and the carbon atom taking hold of the now lonely oxygen atom from the H_2O. For a brief moment the atoms hold on to each other like folk dancers forming an intricate ring—then they burst apart, forming the new constellations. This is what we call a reaction mechanism.

Now the biotechnology kicks in. When manure and urine accumulate in the earth under the cows and horses, tiny nitrifying bacteria (not to be confused with nitrogen-fixing bacteria) get to work on the ammonia molecules. In fact, these bacteria live on them, as their main source of energy is the following reaction (where (aq) signifies dissolved in water):

$$6 NH_3 \text{ (g)} + 12 O_2 \text{ (g)} \rightarrow 6 HNO_3 \text{ (aq)} + 6 H_2O \text{ (l)}$$

This does look much more complicated, and in reality it is even worse as this is a crude approximation of a process containing many different individual reactions.

Confusingly, a chemist might mean two things with a chemical reaction formula. When urea meets water, the reaction

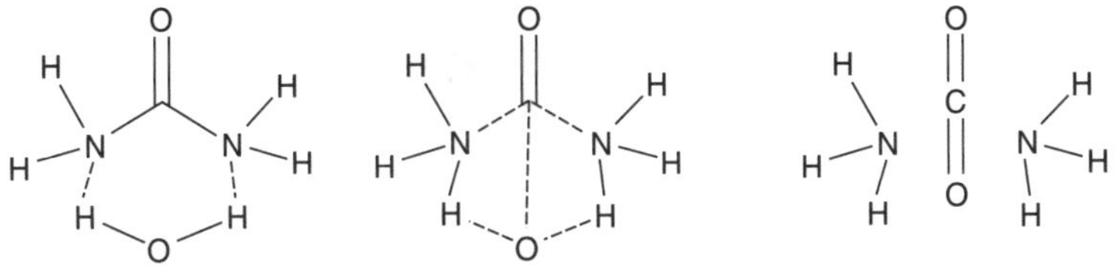

FIGURE 34 A simplified reaction mechanism for the formation of ammonia and carbon dioxide from urea and water. Dotted lines represent new bonds forming or old bonds breaking.

formula describes the actual collision of reactants to give one or two products, in this case ammonia and carbon dioxide. Such reactions normally contain only two molecules, as the probability of three molecules bumping into each other simultaneously is very small—in a similar way, meeting one friend downtown on a Saturday morning may be normal, but having three friends converging on the same spot usually requires some organization, not to mention the 18 needed in the reaction above.

In the nitrification reaction, on the other hand, we look at a whole system: what goes into the bacteria or the collection of different types of bacteria in the soil, and what comes out. This is like young kids entering the school system in one end and coming out as chemical engineers at the other. We know there are a lot of important steps in between, but in a very simplistic way that is how it works. Just as in the school system, where there are (perhaps thankfully) other ways out than becoming a chemical engineer, this formula does not describe what happens to all the nitrogen atoms. However, if we try to make ideal conditions for the bacteria, as was done in the eighteenth century when special saltpetre barns where constructed, we can make a good portion of the nitrogen atoms follow the desired path.

These reactions not only look complicated to us, but are in fact so complex that only a selected few organisms can handle them, and they have evolved a very peculiar piece of molecular machinery to do so. Enzymes are proteins that catalyse chemical reactions in living systems, and they very often carry a metal ion in a strategic place. Some of these you may recognize from the shelf of supplements at the local pharmacy—iron (Fe), zinc (Zn), cobalt (Co), manganese (Mn)—but a nitrifying bacterium not feeling quite up to the job in the morning will find no cure here, because the pharmacist does not normally stock any molybdenum (Mo).

The enzymes used by nitrifying bacteria also contain iron and copper, but the use of molybdenum stands out as an oddity. It is a transition metal, just as iron, chromium, and cobalt are, but comes from the second row of transition metals, not the first, and is thus considerably 'heavier' than the other metal ions used by various living organisms.

But even if the bacteria have done a good job, nitrate and water are not the only things in the petermen's foul-smelling cauldrons: purification is needed. They therefore add the ashes, mostly composed of potassium carbonate (K_2CO_3), also known as potash, when their stew is cooling off. Now we move from biotechnology to the very simple chemistry of common salts—crystalline substances containing positive cations such as sodium +1, and negative anions such as chlorine, which together give what we normally call salt, NaCl(s).

When you put sodium chloride in water, you will notice it disappearing rather fast: it dissolves. This is what normally happens to chloride salts, as they are soluble, but nitrate salts are even more soluble (this is why we don't have mountains of sodium nitrate, as we saw in Chapter 14). The white stuff you get after a barbecue or a campfire, the ashes, is composed of the parts of organic matter that cannot burn or form gases. These are mostly the metal ions, predominantly potassium (K^+) but also sodium, and the carbonate ions (CO_3^{2-}) they need to form a neutral salt.

Adding K_2CO_3 to the cauldrons will produce some white stuff collecting at the bottom. This is calcium and magnesium carbonate, $CaCO_3$ and $MgCO_3$—the same whitish stuff that you find forming in your kettle in areas with hard water. Sodium chloride is also less soluble than KNO_3 and will 'precipitate out' as the chemists say, and both these solids were removed and disposed of. As the cauldrons then slowly cooled, KNO_3 crystals would form, probably first on the surface and on the sides of the vessel, and finally after a weeklong process Per and Esbjörn could normally collect their potassium nitrate crystals and take their product to the King's gunpowder factory for further processing.[131]

$$K^+(hot\ H_2O) + NO_3^-\ (hot\ H_2O) \rightarrow cool\ and\ partly\ evaporate$$
$$\rightarrow KNO_3\ (crystals) + H_2O$$

And then they would move on to the next unhappy farmer. Only on this particular day complications occurred, as buried a couple of feet under Master Jöns' cowshed they discovered a human skull. The sheriff, Ingevald Knutsson Peppanäs, was summoned with his associates and an investigation followed, with thorough questioning of the locals. The extra county court held on 23 July 1708 concluded that the skull had lain there for a long time and that it was impossible to get any further in the investigation, as none of the villagers admitted knowing anything of the matter.[133, 134]

The farmers did not appreciate the petermen, especially as they were sometimes not locals, travelling long distances to do their trade. Could this have been the macabre end of a specially brutal and ruthless member of the force, unwittingly digging his own grave and then being killed by a frustrated farmer? For the record it should be noted that the petermen were not better appreciated in England—for example, their abuses were discussed in parliament in 1606.[135]

An even more titillating possibility is that this has to do with a fall out among the perpetrators of one of the most famous cold cases in Swedish criminal history: the 1676 double robbery of a Swedish Army Treasury transport, known as the 'Loshult coup'.[136, 137] A total of around £4 million in today's currency was taken,* first by farmers at Loshult in Skåne, just across the old border with Denmark, 4 miles south of Älmhult. Then, as the guard and wagons retreated north back into Småland and stopped in Älmhult, just outside the inn, the supposedly loyal Swedish locals stripped the transport of what was left. This is arguably still the largest robbery ever to occur in Sweden, and only a fraction of the booty was ever recovered and only a few of the culprits punished.

The petermen continued their trade well into the nineteenth century. In 1805 the Swedish King ceded his rights to the stable and barn soil, and the potassium nitrate trade was deregulated. In 1830 the government's tax-like demands for a supply of KNO_3 from farmers was abolished, and in 1895 the last remains of the special army agency, the *Salpeterstaten*, was dismantled, as most KNO_3 was now supplied by cheap imports.

Only a few years on, however, both World War I and the use of inorganic fertilizers based on KNO_3 led to another surge in demand for nitrate. This time it was produced not from stinking dung heaps, but by the combination of the colour- and odour-free nitrogen gas (N_2) from the air and hydrogen gas (H_2 obtained from natural gas or coal and water) to give ammonia, followed by

* An accurate estimate of the booty is difficult to make. At least 18,000 daler ('dollars') in silver was lost with an estimated value of £300,000 to £8.5 million depending on method of comparison using *Portalen för historisk statistik—historia i siffror*, <www.historicalstatistics.org/>, a portal for historical statistics, with the main focus on macroeconomic data on Sweden in the nineteenth and twentieth centuries, Rodney Edvinsson, Stockholm University, accessed November 2012.

further oxidation of ammonia to nitrate, via the so-called Haber and Ostwald processes.

Books and plays have been written about Fritz Haber, and he usually stands out as the villain of twentieth-century science, but the petermen, whether villains or virtuous, have been largely forgotten. This is a tribute to them and their stinking trade.

Epilogue

The Greek words *sal* for salt and *petre* for stone is the origin of the word *salpeter* in Swedish and *salpêtre* in French. In English an additional 't' has been added giving saltpetre, and the men of this craft were sometimes called 'salt peter men'. Perhaps an odd coincidence is that the workers of another 'trade' also associated with explosives, safe-crackers, are sometimes called petermen too.

Apart from KNO_3 being produced on a big scale on northern European farms, another chemical mentioned here was also a kind of 'cash crop' for farmers, especially in wood-rich Sweden, namely K_2CO_3, or potash. Its many uses are outside the scope of this note, but one could mention that besides gunpowder it was an important ingredient in a less lethal concoction, that of soap.

17

Bonaparte's Bursting Buttons: A Thin Story

In Chapter 17 we encounter a liquid metal that turns to solid and a solid metal that turns to dust, and in the process we learn more about allotropes, crystal structures, and materials.

On my way to Vilnius, capital of Lithuania, one late November I realized that I had not packed any winter clothes. It turns out that I was not the first to make this blunder. None of the half a million or so Germans, French, Swiss, Poles, Italians, and other nationalities who passed through the town or in its vicinity in June 1812 had packed any winter clothes, something many of them were to later regret.

They were on their way, although they did not know it at the time, to Moscow. What they also did not know was that they were going to make what was arguably the world's worst *aller-retour* journey ever: Vilna to Moscow and back (at that time the town was known under its Polish name and had recently been acquired by the Russians in the process of the annihilation of the Polish state). It was June, and they were in a good mood, as the Russian Tsar had recently fled Vilna followed by his quarrelling generals, and they were under the command of possibly the most competent military leader since Alexander the Great: Napoleon Bonaparte.[138]

The lack of warm clothing was not going to bother me, however. By the morning the snow had melted, and luckily I was not on my way to Moscow on foot. I was in Vilnius to search for some buttons, preferably made of tin.

The story of Napoleon's buttons and their allegedly fateful role in the disastrous 1812 campaign is widespread among scientists and science teachers. This is partly due to the popular book with the same name by the chemists Penny LeCouteur and Jay Burreson,[139] and I wanted to find out whether there could be any truth in it, or whether it was just another of the legends and rumours that has formed around this war.

Briefly, the story goes like this: metallic tin is a dense material (lots of atoms per cubic centimetre) and was supposedly the material used for many of the buttons of what was known as *la Grande Armée*. Unfortunately, metallic tin has a nasty Mr Hyde variation, known as grey tin. This form is nowhere near as dense as metallic (or white) tin, and should part of a button made out of tin metal suddenly transform itself into grey tin, the atoms would need much more space and the button would burst.

This phenomenon is known as *tin pest*, or *tin disease*. The story claims that this is just what happened to tin buttons used by Napoleon's army during the retreat from Moscow in November and December 1812, and that the soldiers therefore could not fight properly, or died of outright exposure as their clothes disintegrated in the extremely cold Russian winter.

Just as diamond and graphite are allotropes of carbon (see Chapter 9), white and grey tin are two allotropes of the element

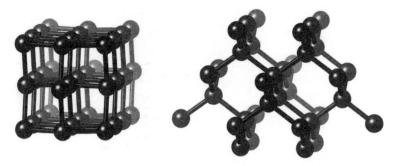

FIGURE 35 Metallic white tin to the left, with its dense packing of atoms, compared to the non-metallic grey tin with the same structure as diamond. The same number of atoms have been plotted in both pictures, and both are to the same scale. See how the atoms in grey tin take up much more space, and imagine what happens to a metallic tin button if it starts to change form. (Faded atoms are further away and the 'bonds' in metallic tin are only to clarify the positions of the atoms.)

Sn (Sn for *stannum*, Latin for tin). Tin sits in the same column as carbon but three steps down. The very idea of the Periodic Table is that elements in the same column should have common properties, and as we see, grey tin conforms nicely to this idea. It has the same structure as diamond, a network built up of two-electron bonds extending in all directions. Consequently we call solid materials of this type network solids: quartz, a network of silicon and oxygen, is another example.

But things also change as we move down in a column, and one effect is that the elements of the carbon group get more and more metal-like. The metals do not have these spoke-like two-electron bonds—they are more like atoms embedded in an electron jelly. How do the atoms arrange themselves when they have no spokes to direct them? In a metal they would like to get as close as possible so as not to create any 'holes' in the electron jelly, and if you want to see a fairly accurate model, just go to your nearest fruit-and-veg

FIGURE 36 A model of the atom arrangement in a metal, the densest way of packing oranges (or atoms) minimizing the free space between them. The seven lighter-coloured atoms to the right show how six atoms exactly fit around one central atom in a plane, a further three above (not shown) and below make the total number of nearest neighbours twelve in a close-packed structure.

shop and have a look. Any self-respecting fruit-monger would stack his or her oranges or apples so that they form an almost perfect model of the atoms in a metal.*

We call this *close packing* in chemical jargon, but if you look carefully at the model of white tin in Figure 35 (the metal form) you will see that it does not exactly conform to the orange-arrangement in Figure 36. This is because even metallic tin has some remaining character of the localized two-electron bonds from the grey, diamond-like form. Metals that conform to the close-packing orthodoxy are, for example, magnesium and zinc.

---

* If the oranges are just deposited in a heap without any finesse whatsoever we will instead get a good model for what we call an *amorphous* solid (without shape). The immediate environment of each orange will resemble the close-packed structure but without any long-range order. For the close-packed ordered structure we can tell the positions of every orange if we know the coordinates for only one, whereas for the amorphous state this is not possible.

The transformation between a metal and a network solid should take place at 13°C, with grey non-metallic tin occurring at lower temperatures, or so thermodynamics tells us. However, thermodynamics can tell us only what changes are possible—it gives no clues about the speed of such changes. In fact, for metallic tin at temperatures slightly below 13°C this reaction is infinitely slow.

You may wonder why there is a metallic form at all if grey tin is more stable and has lower energy. If this arrangement of atoms is more stable at 12°C, then why not at 14°C?

You could think about this in terms of having the most probable system. Take a big container of balls, one of those that small children like to play around in, and fill it to about three-quarters full. Now, a single ball resting on the side will lower its energy and go to a more thermodynamically stable state by falling back down into the basin. In doing so, the total energy will remain the same, the potential energy of the ball having been transformed to heat by hitting the other balls. The reverse process is never going to happen.

However, raise the temperature of the balls by adding small children and the most probable state of the system is going to involve a fair number of balls outside the container. Much the same is going on with the tin atoms, and chemists deal with this by numerically weighing probability, temperature, and the heat generated in the reaction into what we call Gibbs Free Energy. To see if a reaction may occur we calculate the change in Gibbs Free Energy: if that goes down, then the reaction is possible.*

---

* Often, but not always, the reaction that will occur is the one that generates the most heat. The actual formula is $\Delta G = \Delta H - T\Delta S$ where ΔH is the heat generated (a negative number), T is the temperature in Kelvin (always positive) and ΔS is the entropy change (higher entropy being related to 'less order' or higher probability). A reaction is thermodynamically possible when the change in G, ΔG, is a negative number.

The tin transformation starts picking up pace at lower temperatures, with maximum rates reported between −20°C to −40°C, and there are many well-documented cases of tin disease, most prominently in the organ tubes of churches in northern Europe. The Napoleon story is, of course, much more spectacular, and has had a great appeal to teachers and textbook authors who like to spice up their accounts with amusing anecdotes.

So, apocryphal as it may be, in it goes into general chemistry books and other undergraduate texts, usually with some note about its dubious relation to the true story. It also comes in many variations—for example, that it was a matter of cost to use tin instead of brass[140]—and some authors rename battles so that the horrible slaughter at Borodino becomes 'the Siege of Moscow'.[141] Some also take the opportunity to make jokes, painting pictures to the reader of French soldiers trying to fight with one hand while holding their trousers up with the other.

Before coming to Vilnius, I have done my homework, and one thing is clear: this is nothing to joke about. In fact, remarks of this last type are comparable to making fun of the infantry soldiers in Passchendaele or the Somme getting stuck in the muddy trenches during World War I. The events of 1812 were a European tragedy of enormous proportions. About one million people died on both sides, and the final death march from Smolensk in Belarus (where the army thought they would find shelter and provision for the winter, but where there was nothing) to Vilnius makes even Scott's journey back from the South Pole fade in comparison.

It would be easy to think that the information available about these events would be sketchy at best 200 years later, and that we will never know the truth. But on the contrary, there is substantial documentation available, since a large number of survivors, from simple soldiers to generals, wrote down their stories, perhaps as a kind of 'debriefing' before that term was known. Some also kept diaries throughout the campaign, or at least until their ink froze.

One of these was General Louis Joseph Vionnet de Maringoné of the Old Guard, who many years later sat in his chateau close to Gap, where the Alps give way to the hills of Provence, and wrote his memoirs. He can immediately help us with the question of how cold it was on that return journey. He writes that the mercury in his thermometer was frozen solid ('it was like lead') on the morning of 8 December,[142] which puts that night's minimum temperature below the melting point of mercury: −38.8°C. This is in agreement with many other observations at the time, and thus arguably the temperatures were close to the range at which the maximum speed of transformation of tin occurs.

Then there is the question of the buttons themselves: did *la Grande Armée* really have buttons of tin? Was that not too expensive? Some authors say very definitively that it did not, and others claim that we simply do not know.[139,143] Both statements are false. Napoleon's army was very well organized and documented, so of course we know, and yes, they did have tin buttons.[144] Not all soldiers and not for all purposes, but no doubt a fair number of Napoleon's French soldiers had tin buttons on their uniforms when they marched for Moscow. As for the numerous other nationalities in the army, it is less clear, but it seems at least some Dutch regiments were issued with uniforms buttoned up in this way.

This brings us back to Vilnius, and eventually to some anonymous buildings on the east side of the Vilnia River: the Lithuanian national museum's storage and conservation department.

I was there because of a macabre find during construction work on an old Red Army site on the outskirts of Vilnius in 2001.

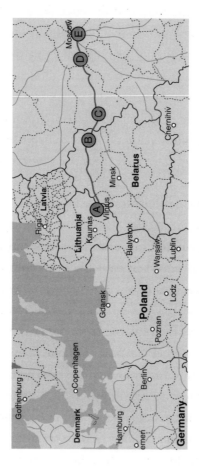

FIGURE 37 The way to Moscow. A map of the marching route of la Grande Armée. In addition, most of the army, except the Poles, had marched a long way even before arriving in Vilnius. (A: Vilnius; B: Vitebsk; C: Smolensk; D: Borodino; E: Moscow.)

When preparing the ground for new buildings, workers discovered a large number of skeletons, and of course immediately alerted the local police. Initially, it was not evident who these poor fellows were, except that they were many: around 7,000. Was this the work of the KGB and the guerrilla war that lasted well into the 1950s? Or World War II? Or the Swedes and Finns marching with Charles XII to their ultimate fate at Poltava in Ukraine in 1709?

Forensic pathologists were called in, and it was soon clear that this was the last remains of *la Grande Armée* of 1812, the poor soldiers and civilian hangers-on (many of these women) that had made it to safety, but died of their injuries in the weeks after.[145] When the Russians re-entered the city these bodies were unceremoniously thrown into the fortifications dug by the French army some nine months earlier.

By a fortunate coincidence, one of these pathologists is a friend of a friend of mine, and that is why we—two chemists, two pathologists, and a conservationist from the museum—were able to rummage through neatly stored cardboard boxes containing the last possessions of this lost army: coat sleeves, hats, helmets, belts and buckles, and, of course, buttons. We found neat and well-preserved brass and silver buttons with the regimental numbers still clearly visible, but only a few, very ugly, tin buttons—some with their chemical analysis attached on a piece of paper with a neat cross in the 'Sn' square.

So, I have seen them, the famous Napoleon's buttons, but if they are ugly because of tin pest or not I do not know, and I thus leave Vilnius richer in experience but only a little wiser.

If this is just an anecdote, who was it that created it, and for what purpose? Or is it a work of literary fiction? At least the most

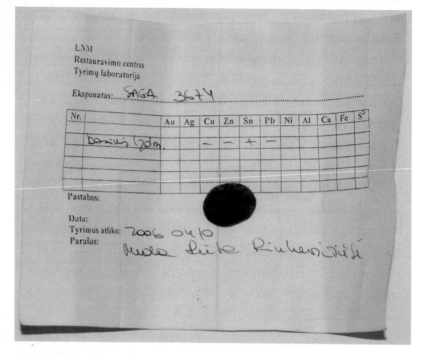

FIGURE 38 One of the famous Napoleonic tin buttons. Black and ugly but mostly intact after almost 200 years buried in a mass grave outside Vilnius, Lithuania.* Photo by the author. © National Museum of Lithuania.

famous author dealing with this war, Leo Tolstoy, seems to be innocent. His famous novel *War and Peace* contains six references to buttons, but none of them bursting and none of tin.[146]

If, on the other hand, this is a true story, and we have established that both the temperature and the material of the buttons were correct, it should appear in at least one of the eyewitness accounts. These are often meticulous and especially macabre, and

---

* In translation: NML (National Museum of Lithuania), Restoration centre, Research laboratory, Exhibit: button 3674, basic alloy, analysis performed by: (signature).

odd details relating to the abnormal cold are related (and often corroborated by other survivors).*

Stories of bursting buttons are, however, conspicuously absent. This in itself does not disprove the theory, as these documents are only a small fraction of the material available. My efforts have to stop here though. I have neither the time nor the skills to go through all these various documents and books in languages such as Russian, Polish, French, and German. Better to ask someone who has been there already, and New York-born, Oxford-educated author and polyglot Adam Zamoyski, of Polish origin, is our man. His recent book *1812: Napoleon's Fatal March on Moscow*[138] was essential background reading for this story, and he tells me he made a point of 'reading every single memoir, letter or reminiscence I could find'.[147] Strangely enough, the story of Napoleon's buttons is news to him, but he confirms my suspicion: there is not a single eyewitness account of pulverized or in any other way problematic buttons.

End of story? Not quite. We can still get an idea of how the rumour started. First, it is clear that this is not a modern invention of an over-imaginative textbook writer. The earliest mention I have found so far is in an issue of the *American Journal of Science* from 1909, and it is clear from the author's phrasing that it was already an established idea by that time.[148] But it was likely that it was already then seen merely as an anecdote. When Professor Ernst Cohen from Utrecht, who became a leading expert on tin disease, gave a lecture on the 'Allotropy of Metals' before the

---

* A fair number of these accounts are also available on the Internet, among them the memoirs of Sergeant Bourgogne and those of our friend with the thermometer, General Vionnet.

Faraday Society in London in 1911, he talked extensively about tin, but did not mention Napoleon.[149]

Instead, he says something else of relevance to our topic: he talks about another button incident reported by the German scientist Carl Fritzsche working in St Petersburg.[150] This took place in an army warehouse during one of the extremely cold winters in the 1860s, and this time it was Russian tin buttons that had disintegrated, not French.

Perhaps someone picked up on this story and jumped to conclusions about the fate of the retreating army in 1812. Not entirely unreasonable, one should add, as we have seen that the weather conditions were suitable (although one may argue about the time needed to convert the metal to powder). There was also important circumstantial evidence in the form of consistent reports on the state of the army's uniforms and general appearance. As I said at the beginning, the soldiers did not have any winter uniforms, and as the cold set in they put on whatever they thought would protect them, including women's dresses looted in Moscow. Also, as their companions fell down dead during the march, or did not wake up in the morning, any useful part of their equipment swiftly changed hands, and especially feet, as good boots were in constant demand. In addition, morale and discipline fell below critical levels in most units and consequently, even before the big chill set in, the ragged remains of la Grande Armée looked more like beggars than soldiers.

As if this was not enough, appalling hygiene and horrific diet, more often than not of raw horsemeat, sometimes cut directly from a living beast, caused many to develop severe diarrhoea. As the temperature dropped, buttoning and unbuttoning became increasingly difficult, painful, and ultimately even dangerous, so as a last resort the soldiers unstitched their trousers, or rather breeches, meaning they cut down the seam from the back waistband, thus making instant relief possible without risking frostbitten fingers and gangrene.

It is easy to see how a superficial reading of the survivors' tales could have led to the conclusion that the disarray of the uniforms was caused by the lack of buttons, or even that the impossibility of buttoning up at −30°C could, by mistranslation or misinterpretation, have led to the wrong conclusion.

For those who still like to believe that the story is true, there is a final straw to hold on to. Thierry Vette, an expert on French uniforms who was on the team investigating the mass graves in Vilnius, concluded that given the number of identifiable uniform remains found, the proportion of tin buttons is very low; a lot of them simply seem to be missing.[151] Giving the general disarray of the clothing and uniforms this is hardly convincing evidence, however.

So, having destroyed this anecdote, can we offer something else to keep our students awake during lectures? Well, there is one very prosaic incident. On the west bound road between Vilnius and Kaunas there is a hill to climb outside the village of Paneriai, and at the time of the retreat this was covered in ice and snow. The lack of a layer of fine particles of silicon dioxide in the form of quartz (more commonly known as sand) on this road prevented the wagons of the French Treasury and other heavy carriages from advancing beyond this point. This caused major financial and armaments losses (although making a fortune for the lucky few who managed to grab sacks of gold from the abandoned convoy before the arrival of the Cossacks). More tragically, the coaches with the sick and wounded also had to be left behind, and many of the patients no doubt ended up in the mass graves uncovered in 2001.[138]

So, the war was not lost for the want of a button, but the treasury was lost for the want of a grain of sand, and perhaps this played a part in the final fall of Napoleon a couple of years later.

18

I Told You So, Said Marcus Vitruvius Pollio

In this chapter we will introduce organometallic chemistry, once the scourge of our cities and now a vital part of drug manufacture and research.

Rural Massachusetts is delightful at the end of summer. The classic New England architecture blends with the lawns, gardens, and green forests into a picture of perfect harmony. It is sunny, and the right time in the afternoon for a stroll around the college town of Amherst. However, after a few blocks a distinct crack appears in this idyll: a traditional white wooden house is being renovated and a skull and crossbones sign on the lawn is telling us to keep out due to danger of lead poisoning.

It turns out that the customary white colour of the houses around here was often due to lead-based pigments. The use of lead in paints was phased out in 1978, but it is still an issue judging from the 16-page pamphlet available in six languages from the US Environmental Protection Agency, and the criminal cases brought against real estate companies and landlords failing to inform tenants and buyers of the lead status of their homes.[152]

Marcus Vitruvius Pollio would probably have agreed with this pamphlet and legislation, and so most certainly would Alice Hamilton. Although almost two millennia separate the Roman engineer from the first woman on the faculty of Harvard Medical School, they are united in the fight against the dangers of lead to the workforce and to the public.

We do not know much about the life of the first century BC architect and engineer Marcus Vitruvius Pollio, otherwise known as Vitruvius, except what can be inferred from his famous work *The Ten Books on Architecture*. This magnum opus, written in the days of the Emperor Augustus, probably represents the summary of the professional experience of an old man. The title is slightly misleading, as architecture in Roman times would cover a much broader area than today. So Vitruvius tells us a great deal about engineering in general, about the chemistry of pigments and, to the benefit of this story, about aqueducts and the proper treatment of water.[153]

He is also clearly a conservative man, lashing out against 'decadent frescos' and 'these days of bad taste'. And no, he is not talking about the type of paintings found on the walls in certain houses of bad reputation in Pompeii—he is simply objecting to the depictions of unrealistic objects, saying: 'pictures which are unlike reality ought not to be approved'.

He is a lover of classic Greek buildings, and his writings were taken up by the architects of the Italian Renaissance, partly because a copy of the book in good condition turned up in the Swiss town of St. Gallen at the right time. So the Italians got the Palladian villas, and a new edition of the book (with added illustrations in part by Andrea Palladio himself) inspired architects for a long time to come.

It is a pity that nobody paid equal attention to the practical advice that Vitruvius gave on lead. When it comes to tap-water arrangements he is absolutely clear: 'water ought by no means to be conducted in lead pipes, if we want to have it wholesome'. Among other things, he cites the bad health of the plumbers, the lead workers (*plumbum* being Latin for lead), as a reason for this as 'the natural colour of the body is replaced by deep pallor'.

As a number of centuries would pass before any Western European city came close to the Roman sophistication in civil engineering, this had no immediate consequences for the general public. However, lead was continuously used for a variety of purposes, and the workers exposed to it (for example printers) suffered the ill effects.

Lead poisoning might, however, have played a significant part in the fate of Sir John Franklin's Northwest Passage expedition of 1845. Franklin's party of 129 men was sent out by the British admiralty in two state-of-the art ships, fitted with steam engines and supplies for three years. Franklin himself was already 59, but had served with merit in the navy and gained fame from exploring the northern parts of Canada.

The ships were sighted by two whalers at the entrance to Lancaster Sound, close to the north-western-most part of Greenland, in the summer of 1845, and then disappeared westwards, into the historic mist of lost expeditions. There has since been an almost continuous string of searches for Franklin and his men, starting in 1847 when the Admiralty in London judged that their food supply had been used up.

There exists a considerable literature, both of fact and fiction, dealing with the Franklin expedition and the subsequent rescue and search missions. Suffice to say that Franklin and his men spent two winters frozen in, north of the polar circle, and as supplies got low abandoned their ships and met their fate on a futile march south.

It might be that technology and the negligence of Vitruvius' advice got the better of Franklin and his men. Remains of various parts of the expedition have been found, and recent analysis of preserved bodies of crew members have shown lead concentrations very much elevated compared to normal levels at the time.[154] There are two possible sources for this lead. It has been suggested that the modern water supply system installed on the ships *Terror* and *Erebus* contained lead parts,[155] and that the tins containing the major supply of provisions, were badly made, exposing the foodstuffs to lead used in the seals.[156]

However, nothing implies that any of the expedition members actually died of lead poisoning. What we do know though, is that one major part of the body injured by high lead intake is the brain, leading to nerve disorder and problems with memory and concentration. Failing to take rational decisions, or even understanding the gravity of their situation, might have led to the disastrous end of the expedition. From excavated evidence, it is clear that many objects brought along on the final and fatal march were just plain silly: books and cutlery just added weight to their burden.[157]

Lead is poisonous in many ways, mostly because it can replace 'good' metal ions such as Ca^{2+}, preventing them from performing their functions. Pb^{2+} is also 'soft', the concept we learnt in Chapter 12, so it will interact with negatively charged soft atoms in the body, and these are typically sulphur, especially in the amino acid cysteine. One of the reasons it causes problems in

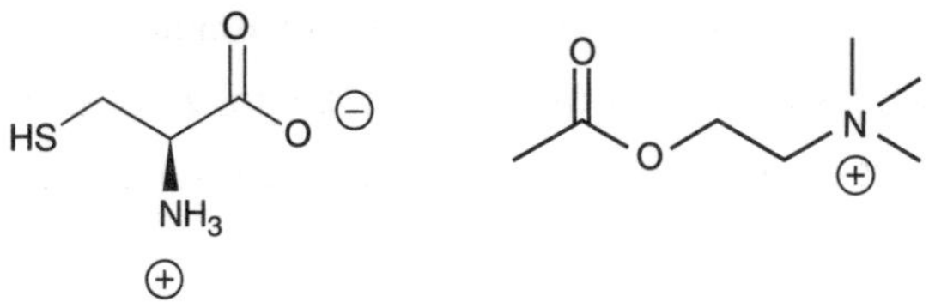

FIGURE 39 The amino acid cysteine (left) and the signal-transmitter molecule acetylcholine (right).

the brain is because it interferes with enzymes regulating the neurotransmitters acetylcholine and dopamine.[158] The 'deep pallor' observed by Vitruvius, on the other hand, comes from its effects on the haemoglobin system.*

Severe mental illness from lead poisoning was the reason Joseph G. Leslie was locked up in a psychiatric hospital for 40 years. When he finally passed away in 1964, it was to the great surprise of most of his family as they believed he had died already in 1924 in an accident at Standard Oil of New Jersey, where he was a chemical operator. Only his wife and son knew the truth.[159]

Although work conditions were frequently dangerous in those days, the accident, which involved several workers, got quite some publicity. This is where Alice Hamilton comes into the story. A physician educated at the medical school at the University of Michigan, she was one of the pioneers of occupational health in the US. In 1919 Hamilton was just starting her work on the faculty at the Harvard Medical School's department of Industrial Medicine. In the same year, General Motors in Detroit acquired a small research company run by a Charles Kettering in Dayton, Ohio.

---

* It should be noted that it may matter a great deal, both for the acute toxicity and the ease of intake, in which form or in which kind of molecule the lead is present. For example, metallic lead is more difficult to get into your body than tetraethyl lead, which is a liquid.

Kettering was hunting for the perfect anti-knocking agent for car engines, and in 1921 his team came up with tetraethyl lead as one possible solution.[160, 161]

This time, however, the public and civil authorities were aware of the potential dangers of lead. Furthermore, as neither manufacturer, E.I. du Pont de Nemours Corp nor Standard Oil, could master the tricky production process (in addition to the non-death of Joseph G. Leslie, ten other workers died and more were injured from lead poisoning), the future of the new additive looked gloomy. Alice Hamilton, who was becoming an authority on the subject of industrial toxic substances, spoke out against 'ethyl', as the proponents in the Ethyl Gasoline Corporation liked to call the substance. Hamilton stressed the risk both to the safety of workers and, in the long term, to the entire population. Tetraethyl lead was prohibited in New Jersey and no longer sold in New York and Pennsylvania.

Although Kettering had other possible solutions up his lab-coat sleeve, it seems General Motors, and Standard Oil (the principal stakeholders in Ethyl Gasoline Corporation) decided that tetraethyl lead would be the optimal commercial choice. The patent situation was clear, and in a situation in which 'ethyl' could become a standard addition to any gasoline, profits were going to be substantial. This was certainly not the case with ethanol, another alternative, or iron carbonyl, where there were European competitors and possible patents.

Chemically speaking, iron carbonyl and tetraethyl lead (shown in Figure 40) are known as *organometallic compounds* because they contain a metal–carbon bond. Tetraethyl lead is rather straightforward when it comes to chemical bonding. We can think of lead behaving a bit like carbon, the element at the top of that particular

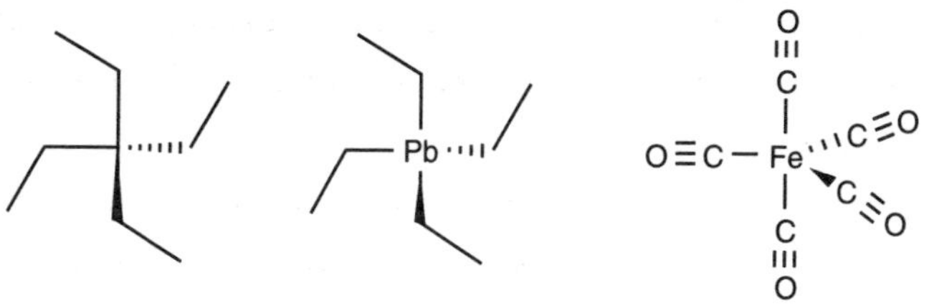

FIGURE 40 A simple hydrocarbon, the lead derivative tetraethyl lead, and iron carbonyl, $Fe(CO)_5$.

column of the periodic table, so that $Pb(CH_2CH_3)_4$ would be an analogue to the hydrocarbon $C(CH_2CH_3)_4$, 3, 3-diethylpentane shown in Figure 40. Iron carbonyl or $Fe(CO)_5$ is a different matter altogether.

If we think about the C–C bonds in 3, 3-diethylpentane as composed of one electron from each carbon shared equally by the two atoms, we can think of the Pb–C bond as a Pb^{4+} ion borrowing two electrons from each negatively charged ethyl group to form four Pb–C bonds, also with two electrons in each. Fe in $Fe(CO)_5$ is neutral, as is CO or carbon monoxide, so in this case it is not obvious in which direction the electrons will want to move. It turns out they move both ways: two electrons at the end of the CO molecules are shared with the iron at the same time as other electrons, stabled in a different orbital, are given from iron to CO. This is so complicated that we usually do not try to show this when way we draw the bonds, so there is still only a single line connecting Fe and C, although the character of the bonding interaction is completely different from the Pb–C bond.

Whether $Fe(CO)_5$ would have been a better anti-knock agent than $Pb(CH_2CH_3)_4$ is hard to say. Their acute toxicity seems to be about the same, and perhaps you would need more iron carbonyl added to the fuel. Anyway, this does seem to show that an alternative that did not spread persistent pollutants could have been developed as early as the 1920s.

However, few people knew about this at the time. To resolve the question of tetraethyl lead, a national meeting was called by the US Surgeon General, and in May 1925 over 100 delegates from companies, trade unions, government agencies, universities, and the press met in Washington DC. Tetraethyl lead was presented by Kettering, one of the main speakers, as the only viable alternative to solve the knocking problem, although some, including Hamilton, refused to believe this.[162]

Of course, Kettering's alternatives were securely locked up in his corporate labs—we only know about these through the work of Professor William Kovarik at Radford University, who has used recently released files* from GM's research organization in Dayton to write a comprehensive study of the case. The pressure from competing products was probably also the reason why GM and its associates started the large-scale production of tetraethyl lead before developing adequate safety procedures and properly training their workforce.

The conclusion of the meeting was the formation of a committee, and this committee finally found that it had little legislative power to forbid tetraethyl lead. Vague promises of further investigations were forgotten, and by the end of the 1930s 90 per cent of all US gasoline was 'leaded'.[163]

Still, in a way it was a victory for Hamilton and her colleagues, as the proceedings created a precedent for the right of government and science to have the last word in questions of industrial hygiene and pollution.

---

* These files are deposited at Kettering University in Flint, Michigan.

Vitruvius was probably a lonely voice in an age in which his fellow Romans happily spiced their food with lead(II) acetate trihydrate,* and later on *The Ten Books of Architecture* was probably not required reading for mining engineers, metallurgists, and printers. That Vitruvius' advice was neglected may perhaps be excused, but that the heated lead debate of the 1920s was completely forgotten is more remarkable.[†] Kovarik points out that in a 1984 *New York Times* report[‡] on the then recently introduced Chicago prohibition of tetraethyl lead, this is hailed as the first of its kind. If the reporters had searched their own archives they would have found similar reports on the 'ethyl' bans 60 years earlier.

Scientists too seem to have committed similar errors. In a 1925 article in the widely respected *Journal of the American Medical Association*, Hamilton and her colleagues give a comprehensive overview of the toxicology of tetraethyl lead and completely trash an earlier Bureau of Mines report that plays down the effects of lead-tainted exhaust gases.[164] This article has been cited only 15 times by other scientists, compared to the many hundreds of articles dealing with the issue.

On a final note, we see that Kettering had a university named after him, but every year the US National Institute of Occupational

---

* Also known as *lead sugar*, this compound could be thought of as a 'traditional, all natural alternative to sugar', but is of course as toxic as other lead compounds.

[†] Aaron Ihde notes that after the Surgeon General's Report in the 1920s: 'Despite the use of tetraethyl lead in almost all motor fuel and despite the proliferation of automobiles there has never been another medical study of the effects on human health.' *The Development of Modern Chemistry*. Ihde, A. J., Dover Publications, 1964, 1984, p. 710.

[‡] 'Chicago issues a ban on selling leaded gas.' *New York Times*, 8 September, 1984, <www.nytimes.com/1984/09/08/us/chicago-issues-a-ban-on-selling-leaded-gas.html>

Safety and Health (NIOSH) gives out the Alice Hamilton awards for Occupational Safety and Health. NIOSH state on their homepage that:

> "Many of the first laws and regulations passed to improve the health of workers were the direct result of the work of one dedicated and talented woman, Alice Hamilton, M.D."[165]

Hamilton lived to be 101, and died just as the second heated tetraethyl lead debate started in the 1970s.

Although now phased out in most of the world, the controversy surrounding the introduction of tetraethyl lead seems to go on. Could GM, or other companies, have developed a viable alternative in the 1920s? We will never know, but perhaps the brilliant scientist Thomas Midgley, Jr, who lead the anti-knocking research at GM, could have been known as the man who really solved the knocking problem, and not, sadly, as the man who introduced one of the most persistent environmental problems of the twentieth century.

What we do know, however, is that the early work on compounds like tetraethyl lead and iron carbonyl paved the way for a small revolution in organic synthesis: the introduction of reagents and catalysts based on organometallic compounds. Although neither $Fe(CO)_5$ nor $Pb(CH_2CH_3)_4$ is used directly as a reagent or catalyst, the understanding of these dangerous molecules has, somewhat ironically perhaps, given us better and more environmentally friendly routes to many chemicals and drugs—something we will learn more about in Chapter 22.

19

A Shiny Surface and a Tainted Past

*In this chapter we will learn about an element with good and bad
sides, oxidation and prevention of oxidation, the fast way to make
vinegar from ethanol, and we also use a hound to shepherd some
zebras.*

The day Erin Brockovich was driving in Reno and got hit by
another driver, brought her in close contact not only with the
bumper of the other car, but eventually also with the US legal
system, and this would change her life completely.

The day Steven Soderbergh asked Julia Roberts to play the part
of Erin Brockovich in the film with the same name didn't really
change her life, one presumes, but it would show the world's movie-
goers and critics that the star and Academy Award winning actress
of 1990 was really back on the right track.

What is the link between these events? The answer is the element
chromium.

It was chromium that made law-firm clerk Brockovich start a
David-against-Goliath struggle with the California energy conglomerate Pacific Gas and Electric Company, that made director

Soderbergh make the blockbuster movie that gave Roberts an Oscar for best female actress in 2000 and revitalized her career.

I will try not to spoil the picture for those who have not seen it, because it is well worth watching, but the fact that the good guys win in the end is probably not a surprise anyway. However, the role of chromium in this play is not at all evident. And are the good guys really the good guys?

There is usually a proper amount of, and a proper place for, everything, and this includes the elements of the periodic table. The main component in steel, a material which has a role to play in this story, is iron, and while we sometimes have too low a level of this element in our bodies, too much of it will kill us.

The same goes for chromium: we can't live without it. Or so it was thought until very recently.[166] It was supposed to help us to break down and metabolize sugars, and thus 'chromium deficiency' could possibly be related to diabetes.[167] Now, while low levels seems to do no harm, there are still possibilities of a *therapeutic window*—that is, concentrations where it may do some good—but it does not any longer seem to be considered an essential element, although official consensus on this has not yet been proclaimed.[168] What is clear, however, is that too high a level of chromium causes problems, and the form in which it is digested or inhaled is significant as well.

In Hinkley, California, chromium was neither present at the right time (the 1980s), in the right place (it was in the drinking water), or in the right amount. What's more, it was disguised as the chromate ion, CrO_4^{2-}, a sort of Trojan horse that efficiently hides its toxic contents inside a tetrahedron of oxygens. In this form, the guards at the walls of our cells will let it pass, believing it is a harmless lookalike: the sulphate ion, SO_4^{2-} (see for yourselves in Figure 41).

How did it get in the drinking water? As a result of a simple engineering mistake, or cutting corners to save costs? We will probably never know. In any case, PG&E, as they are known, were pumping natural gas around California as part of their energy infrastructure. Over long distances such pipelines need repressurizing at intermediate compressor stations, and one such station was situated in Hinkley.[169]

Even though the pipelines themselves are not so large (with diameters of 24–26 inches (61–66 cm)), and buried underground, a compressor station is a sizable industrial installation, with large parts of the equipment, tanks, and tubes made out of steel. Steel is prone to corrosion, and therefore the water in the water tanks used on the premises were spiced up with a corrosion inhibitor.

While corrosion may seem like a minor inconvenience, involving some dirty work on your car now and then or being ripped-off

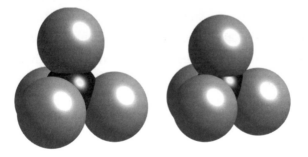

FIGURE 41 The evil chromate ion, CrO_4^{2-} to the left and its benign lookalike the sulphate ion, SO_4^{2-} to the right. Pictures are drawn to the same scale. This style of drawing emphasizes the overall shape and 'electron density' of the molecules and not the chemical bonds. The Cr–O and S–O bonds are different, but these are not 'seen' from the outside—an approaching molecule would only sense the electron density from the oxygens.

at your local garage, at industrial installations and major parts of our infrastructure such as bridges and railways, corrosion costs incredible amounts of money every year. So an important part of all engineering projects is to fight corrosion, and one way is, or was, to use the chromate ions. These will stick to the steel surfaces and form a thin but efficient layer somewhat resembling that formed on stainless steel (which also contains chromium), keeping nasty oxidants like oxygen away from the iron. Making such thin 'passivating' surfaces is a successful anti-corrosion strategy: we have seen it with aluminium in Chapter 10. For aluminium, the natural oxide coating is very efficient at corrosion protection, unlike iron and steel, which need chemical manipulation.

The use of chromate ions to prevent corrosion is not necessarily a problem—we can learn how to deal with dangerous things in safe ways (electricity is a prime example). The problem was that PG&E did not deal with it at all. Instead of installing some simple devices protecting the surrounding environment, the chromate-tainted water was leaching out of the installation.

This was going on for many years, and while rummaging through a filing cabinet in the law firm where she worked Brockovich found a disturbing number of illnesses in the Hinkley community and persuaded her boss, lawyer Edward Masry, to be allowed to investigate. After a lot of footwork on Brockovich's part, and legal manoeuvring by Masry, they won the biggest lawsuit ever against PG&E. The final settlement was $333 million dollars on behalf of members of the Hinkley community.[170]

So far we are following the film script. The bad guys got their punishment and the heroes their reward. However, some people would argue that the health of the general public would be much better served if more attention was given to real risks, like tobacco and alcohol, and less attention was given to perceived, but scientifically non-proven, risks like trace amounts of pesticides in vegetables. Or indeed chromium(VI) in drinking water.

Some would even say there was no *scientifically proven risk* with the chromate levels found in the Hinkley drinking water,[171] and that the whole affair was just costing PG&E customers and shareholders a lot of money.[170, 172]

It turns out that at the time of the legal proceedings, the dangers of these levels of chromate ions in drinking water was far from proven. This was partly due to lack of data. Chromate ions in water had not been perceived as a problem before, and there was not much research to argue about.[173] It was, however well known that chromate-containing particles cause lung cancer,[174] and American unions had some years earlier taken the US Occupational Health Agency to court because of its alleged tardiness in lowering the permitted limits of chromate in the air: limits based on solid scientific results.[175]

Moreover, chromate ions and Cr(VI) have been known as carcinogenic and mutagenic since the 1950s,[176] and no engineer or chemist in their right mind could have thought it a good idea just to let these compounds seep down into the groundwater.

Part of the problem was probably also that the chromium procedure was no longer in effect at the Hinkley station. It had already been discontinued in 1966, and the chromium-tainted water had then slowly crept down from leaky ponds to contaminate the groundwater many years later. That such a long period passed before the problem surfaced may have meant that PG&E management felt less of a personal responsibility, as this was the mistake of an earlier generation of management.

So, at least according to the movie, they tried to dodge the problem, and even explained to the Hinkley inhabitants how good chromium was for you (and technically of course they were right: as we have seen, at that time chromium was considered to be essential to humans). This did not, however, convince the judge, even though PG&E found expert witnesses to testify for them (one of whom was later to be appointed by the George W. Bush administration to an advisory committee of the National Center for Environmental Health).[177]

Proving beyond doubt that diseases developing over a long period of time are caused by the digestion of low levels of toxic compounds is very tricky, but this is hardly needed here. What we have is a company that, by neglecting best practice engineering, contaminates the groundwater with a molecule with well-known toxic properties. The concentrations of this compound then exceed the levels set by the authorities,[178] and of course punitive measures are in place, irrespective of whether someone actually got sick as a result or not. If you drive over the speed limit, try arguing with the police that it is OK, just because so far you have not killed anyone. Whether the settlement of $333 million dollars was appropriate is another matter entirely.

But let us end this chapter on a more chemical note: the oxidation state. The movie is full of references to 'chromium-six', which is chromium in oxidation state +6, which is often written with Roman numerals as chromium(VI). This is also called hexavalent chromium, implying that the metal ion can form six bonds to other atoms, which is in fact an archaic and not very helpful idea, as the small Cr^{6+} ion normally does not bond with more than four other atoms because otherwise it gets too crowded.

Referring back to the Preamble, making Cr(VI) means removing all the electrons in the last filled *s* and *d* orbitals (or pens if we prefer the zebra allegory) from the chromium atom so that, in terms of the number of electrons, it becomes just like the noble gas argon. This likeness to unreactive argon lends some labile stability to this very high oxidation state, compared to chromium(V) for example—like balancing a coin on its edge. It is a very reactive oxidation reagent though, since it does like to fill some of these empty pens with electrons again, and in the presence of something that may give up electrons the coin may easily fall over.

This is also the reason that drinking water with very low concentrations of chromate is seen as less of a risk than inhaling chromate particles (or indeed water vapour containing them). The chromium(VI) should be reduced in the body before it has time to do any harm.

We can use chromium(VI) to make carboxylic acids from alcohols in a much more rapid and efficient way than just leaving a bottle of wine open and letting oxygen in the air produce vinegar. The chromium will then go back to a more stable oxidation state, chromium(III). For the positive ions of chromium and all other transition metals, the *s*-orbitals are now less attractive (they are higher in energy) and only the *d*-orbitals can be filled with electrons. The question is: how do we now arrange the electrons in these available orbitals?

It is much better to have one electron alone in one orbital than to have two together, as they will repel each other. It is also better to let them have the same spin (or, as we said before, the same type of stripes on the zebras), as when they reach the water hole (the nucleus) they will come close to each other and having the same spin will make them automatically avoid each other. So we will have one electron each with the same spin in three of the five d-orbitals, and this is known to chemists as Hund's Rule.

Why do we normally get oxidation state +III and not +IV or +II when Cr(VI) is reduced? This is because these ions will attract six other atoms around themselves, Cr +II, +III, and +IV being larger than Cr +VI. We exemplify this with the $[Cr(H_2O)_6]^{3+}$ ion shown in Figure 42 and see that this looks something like an octahedron.

When you do this, it turns out that three of the five different d-orbitals become more stable, because we tend to let in pairs of electrons from the surrounding molecules or atoms in the other two (there will be a lot of competition for the grass in these). The best arrangement you can have is then one electron in each of the three more stable orbitals, leaving the other two empty, symbolized by the arrows in the little boxes in Figure 43. Counting the electrons we see that this is just what happens for the chromium(III) ion, and this is therefore a nice and stable oxidation state.

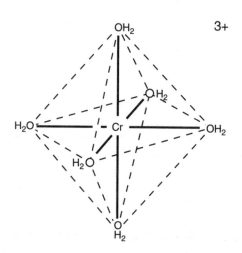

FIGURE 42 The $[Cr(H_2O)_6]^{3+}$ ion shown with the Cr–O bonds in bold and the octahedral coordination geometry emphasized with dashed lines.

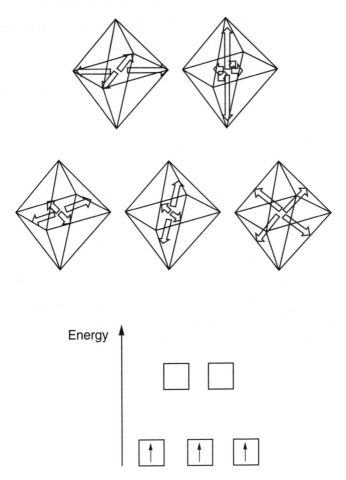

FIGURE 43 The principal directions of the five d-orbitals available for the three d-electrons left in the $[Cr(H_2O)_6]^{3+}$ ion. Two of them are directed towards the oxygen atoms in water and prefer to accommodate electron density from these oxygens. This leaves the three other orbitals for the three chromium electrons, depicted as arrows in the little boxes of the orbital energy diagram at the bottom of the picture.

After the car accident in Reno, Erin Brockovich and the driver of the other car may have been subject to a quick chromium(VI) test. The possibility of alcohol consumption being the cause of an accident is sometimes considered, and at that time the simplest test was to see if the breath of the driver would make the bright red-orange colour of chromium(VI) transform to the less intense green colour of chromium(III) when Cr(VI) reacted with ethanol in the breath to form acetic acid and Cr(III) ions.

20

The Actress and the Spin Doctor

We learn about Nuclear Magnetic Resonance (NMR), the preferred way of chemists to identify new drug molecules and for doctors to diagnose brain tumours. We also meet a rare element that is not so uncommon after all.

This story could begin with a fictional horse named Velvet, or perhaps behind the keyboards in a smoky bar in Halmstad, and it might have had a very different ending were it not for the unlikely excursion of a few medicinal chemists to the exotic southern part of the periodic table into the realm of the Lanthanoids.

One or two generations of movie aficionados and the celebrity-hungry populace were upset, intrigued, or just plain nosey when another chapter in what seemed to be the never-ending story of Elizabeth Taylor, last of the great Hollywood divas, was revealed in February 1997. Dame Elizabeth had been diagnosed with a brain tumour just before her 65th birthday and was due for an operation in a few weeks.[179] As it turned out, this non-malignant tumour, which was easily removed by surgery, was probably the least of her medical problems, but when singer Marie Fredriksson fainted in her home in 2002 and the causes were revealed to be a potentially deadly cancer tumour, the situation was radically different. The fate of the 44-year-old, half of the immensely successful pop duo Roxette, affected a different generation to those who had been following the career of Elizabeth Taylor since the 1940s, and was also deeply unsettling since the victim was a woman in the prime of her life with young children.[180]

Brain tumours are difficult to deal with—you cannot just break open the skull and poke around until you find them, there far too many sensitive connections and devices that may be broken. Their precise location is a key issue, and for that you need to look inside the head without opening it. The most powerful method is what scientists call *Nuclear Magnetic Resonance Imaging*, known to the public simply as *MRI* as the little word 'nuclear' may have associations that would be unhelpful in situations in which calm and composure may be necessary, both for the patient and relatives.

Looking inside human bodies has been of tremendous importance for medicine, but has also meant business opportunities, both for the unscrupulous and the upstanding. Burke and Hare sold their murder victims as exhibits for student demonstrations at the Edinburgh Medical School, a business slightly more risky than that of Sweeny Todd, as someone was bound to recognize one of the bodies eventually. On the right side of the law, the 2003 Nobel Prize laureates in medicine, Paul Lauterbur and Sir Peter Mansfield, made some of the most important developments for the MRI technique.

Many of us who have children have seen ultrasound images of our yet-to-be-born offspring, which is a moving moment, but frankly the resolution is not that great. I remember seeing a movie of a living, beating heart inside some animal (I have forgotten which species) at a Nuclear Magnetic Resonance conference in Switzerland in 1991, with a resolution that made the ultrasound image look like a 1980s computer screen compared to a 2013 laptop display.

I will get a bit more technical in this chapter than in the preceding ones, partly because I spent most of my five years as a PhD student working with a Nuclear Magnetic Resonance spectrometer, and partly because this is such an important diagnostic tool. Anybody who habitually watches one of the ubiquitous hospital drama series on television has most likely seen the patient-of-the-episode loaded into a full-body MRI instrument. I can assure you that you will get out alive from such an experience, although it does look a bit like science fiction, but for anyone slightly nervous about machines and electronics, especially since I've mentioned the little word 'nuclear', it may be comforting to know how it works.

For the rest of us, it is just fascinating. My sister-in-law was once approached in a hospital by a somewhat worried looking nurse asking her if 'she thought there were too many machines' around her, and she answered, 'No, keep 'em coming, I'm an engineer!' I would like to start this with a tribute to the often forgotten engineers of this world. Somehow it seems to have a higher status to know the theory than to know how the machine works. For example, to be able to explain what goes on when we put a water molecule in a magnetic field and hit its hydrogen atoms with a radio frequency and then 'listen' to the signal they send back to us is highly regarded. However, to understand how we keep the magnetic field so extraordinarily stable that the trams moving on Universitätsstrasse in Zürich interfered with the NMR spectrometers at the ETH (Eidgenössische Technische Hochschule, the Swiss Federal Institute of Technology) a block away, how we pick up the extremely weak radio signal from the sample and transform it into the nice signal we expect from 'knowing' how the method works, and how the analogue-to-digital transformer makes the signal processable by computer, to mention a few examples, are to many no more than technicalities.

I will shortly fall into that trap too, but I must say that sometimes I find that making these machines work at all, and turning them out in large numbers from factories, is as amazing as the underlying science. I agree that for many people, knowing the principles behind these technologies is more important. It is obvious to anyone that the big boxes surrounding these machines are not empty but filled with 'electronics', and it is more important to understand that 'nuclear' in nuclear magnetic resonance does not imply any kind of 'radioactive' or ionizing radiation. However, for the chemist working behind the controls of an MRI spectrometer an understanding of the electrical engineering may be more important than knowing the detailed quantum mechanics worked by operator algebra and Bloch equations, although in the scientific community the latter is far more prestigious.

So, what is actually going on inside the MRI machine? Sometime during the late 1930s physicists discovered that the nuclei of some isotopes, for example the normal ^1H isotope of hydrogen and the very rare ^{13}C isotope of carbon, behaved as tiny little magnets. They set out to measure how strong these magnets were (their 'magnetic moment') using large electromagnets, but as there are only a fixed number of isotopes, soon all the magnetic nuclear moments had been catalogued and, by the end of the 1940s, this was a closed chapter—or so they thought.

Then somebody realized that the magnetic field around the nuclei is not exactly the magnetic field generated by the electromagnets used in the measurements. It is slightly altered by the electrons around the nucleus, which have their own tiny magnetic moments associated with their spin. As some of these are involved in chemical bonding, or may even have been lost to another atom forming an ion, this will further modulate the magnetic field. The result is that the energy needed to switch the small nuclear magnets from pointing in the direction of the external field (just as compass needles point in the direction of the Earth's magnetic field) to point in the opposite direction will change slightly. This change only affects the fifth decimal place in the radio frequency waves we send into the material (or patient), and that supplies the energy needed. It is, however, enough—provided the engineers have done their work properly—to make the signals from a hydrogen molecule attached to a oxygen atom in water different from a hydrogen atom bound to a carbon atom in a protein.

Energy levels, such as for a hydrogen nucleus with only two options available—what we, by an apparently not very good analogue with the world of big solid objects, call 'spin up' or 'spin down'—are central to quantum mechanics.* Forget things like Heisenberg's uncertainty principle and Einstein's famous 'God does not throw dice' opposition to certain parts of the theory—without the quantization of energy levels we would not have been able to observe Elizabeth Taylor's brain tumour, or anything else for that matter. Green, blue, and red colours are also an effect of quantum mechanics. The world would be a grey place if the energy levels of atoms and molecules were not quantized, taking discrete values rather than forming a continuous band of energy,

---

* Where 'quantum' means that there are specific energy levels, like in a ladder, that are possible, not just any energy.

because this structure enables us to see distinctive colours, not just shades of grey.

Chemist use NMR to verify the synthesis of the active ingredients of Tamiflu or aspirin by dissolving the compound in a solvent and checking that all the appropriate signals are present for the ^1H and ^{13}C isotopes. This is the way specialists in organic synthesis and medicinal chemistry work when they analyse promising new molecules isolated from nature, synthetic modifications of these, and completely new man-made compounds.

To observe a brain tumour, we need to apply the technique in a slightly different way. For a certain strength of the magnetic field, the water molecules present in all parts of our bodies will need a certain energy, a radio pulse, to be pushed up to the next quantum level. This is called the energy of resonance, and what we then actually detect is the radio signal sent out again when the nuclei drop back to base camp. The stronger this signal is, the more water molecules are present.

The trick applied in the MRI machine is to have a different magnetic field* for every part of the body. Imagine dividing the body into tiny little cubes, with a different magnetic field in every cube. Every cube will then send us back a specific radio signal, and the strength of this signal will show us how much water that specific cube contains. However, now comes the real twist of the method—all organs, or indeed tumours, have slightly different water concentrations, so the image we can construct (on the condition that we can keep track of the strength of the magnetic field in each cube) will show the inner workings of the body.

This seems a long way from the southern parts of the periodic table, where we find the element gadolinium—little known outside

---

* These fields are generated by helium-cooled superconductors; one reason not to buy helium balloons for the kids!

the area of inorganic chemists and, as it turns out, MRI specialists. There could be a number of places where an exotic element could turn up in such a complicated device—in the semiconductor devices, in the magnets, in the radio receivers, or in the analogue-to-digital converters. Surprisingly enough, this rather toxic metal makes its appearance as a contrast agent: a compound that will make the pictures of our insides, especially the brain, much clearer and easier for the physicians to interpret.

If you have been unlucky enough to have had severe problems with your stomach tubing, you may have had a normal X-ray photo taken, preluded with a barium meal: a preparation of the very insoluble barium sulphate. These barium atoms will absorb the X-ray photons and thus prevent them from hitting the photographic film, or these days an image plate (briefly discussed in Chapter 7), and thus give clearer black and white images of the intestines. One would expect the gadolinium ions to absorb the radio waves, but although the function is the same—to give a better contrast between organs where the agent is present and the surrounding tissues— barium and gadolinium work in completely different ways.

We can imagine the nuclear spins, the small nuclear magnets, as wind-up toy cars. The ground state of the toy is like an unwound device parked in the toy box. The excited state, as we like to call the high-energy state, is the wound up but not-yet-released car. As you let go of the key, the mainspring is released and the little toy speeds around making a noise. The more cars, the more noise, and

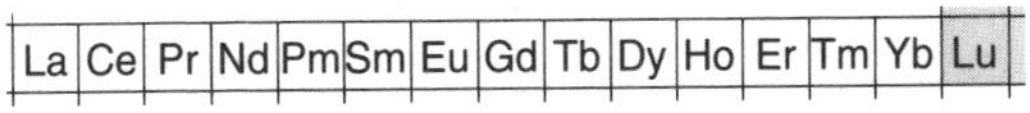

| La | Ce | Pr | Nd | Pm | Sm | Eu | Gd | Tb | Dy | Ho | Er | Tm | Yb | Lu |

FIGURE 44 The southern part of the Periodic Table—the realm of the Lanthanoids.

it is the noise we record: this corresponds to the MRI radio signal. Doing this just once will create a very low signal, so we need to rewind the toys, let them go again and add the signals together, and keep repeating this until we get a reasonable picture.

This may sound uncomplicated, but the problem is that we need to wait quite some time until the little car stops with the spring completely relaxed, so that we can wind it up again. This is where the gadolinium comes in, in the form of Gd^{3+} ions. The ions we knew from school usually had all their electrons in pairs. The ubiquitous sodium +1 ion has ten paired electrons, the negative chloride ion has nine electron-pairs (and thus 18 electrons in total). By contrast, the Gd^{3+} ion has seven unpaired electrons in its outer electron shell (and an additional 54 in pairs closer to the nucleus), which is something of a record. As an unpaired electron behaves like a tiny magnet, in the same way as the ^1H nucleus, it is easy to imagine that this will somehow affect the water molecules.

On condition that the water molecules get close enough to the Gd^{3+} ion, the effect on the excited ^1H nuclei is dramatic indeed—the unpaired electron's magnetic moment makes the H nucleus drop down from the excited state more quickly, corresponding to lifting the wound-up toys and unleashing the spring without resistance, making a big noise in a very short time. Now we can almost immediately hit the patient with another radio pulse, winding up the ^1H nuclei in the water, and again collecting data very quickly. The result is dramatically improved pictures, a shorter time for the patient in the MRI machine, and many more patients 'scanned' in a day.

What about the toxicity of the gadolinium ions—should we not be concerned about that? The truth is that we are, although perhaps chemists less than physicians. 'Real' doctors are normally very reluctant to inject their patients with metal ions that have no known positive role in human biochemistry, perhaps because of the medical profession's early tampering in this area with, for example, the widespread and mostly detrimental use of various mercury (Hg) preparations against syphilis. However, if we take a look at how much of various metal salts are needed to kill a rat (on average, known as the LD50 value) it turns out that you need almost twice as much gadolinium nitrate as you need potassium nitrate (Chapter 16), more than 5 grams per kilo of rat (that may seem like a very big rat, but is just the unit we use for these measurements).[181, 182]

This does not mean there are no concerns. Some gadolinium contrast agents are not recommended for people with kidney problems,[183] but this may be more due to the big wrap-around organic molecules that engulf the gadolinium ion. These molecules, looking much like a squid in the process of consuming its prey, are called many-teethed-binders, or rather (as chemists do like to talk Greek) *polydentate ligands*, a ligand being a single molecule attached to a metal ion. They fulfil both the purpose of chaperoning the metal ion out of the body via the urine, and guiding the gadolinium ion to specific organs that needs imaging enhancement.

These molecules normally have a small opening (illustrated in Figure 45), leaving room for a water molecule from the body fluids to make close contact with the gadolinium ion and rapidly unwind, or as we say, relax, and be ready for the next radio pulse.

In rare cases, and for some contrast agents, problems occur, but in general this is a very safe procedure. The organic molecules bind the gadolinium ion using a principle known as the *chelate*

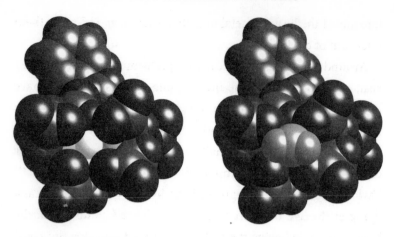

FIGURE 45 In gadolinium contrast agents such as this, the metal ion in the centre is almost completely surrounded by a large organic molecule, left, leaving a small gap for a water molecule (light grey) from the body fluids to make close contact and 'get relaxed' (i.e. drop down from the excited state), as in the right picture.

effect—another word derived directly from Greek* and meaning 'claw'. These polydentate ligands are of the same kind as the antidotes employed in cases of metal poisoning. Apparently, television screenwriters have an unlimited imagination when it comes to making fictional patients inhale, drink, eat, or inject various not-too-good-for-you metal ions, and then when the doctor has worked out an episode's medical riddle he or she triumphantly orders the patient to be put on *chelation therapy*. This therapy means the intake of 'empty' polydentate ligands, which on their way through the organs will hungrily put their claws (or teeth)

* In passing I have noted that students and practitioners of the natural sciences are, in these days when classical Greek has all but disappeared from the arts and letters curriculum even in the UK, among the very few having some kind of classic Greek vocabulary at all, and who recognize that 'Λ' is the letter 'L' and not some kind of weird Greek 'A', and that 'Σ' is an 'S' and not an elaborate 'E'.

into any of the heavier metal ions they encounter and carry them safely out of the body.

Around 60 million MRI scans are performed worldwide every year, many of these of the brain, using contrast agents. I have not had access to the personal files of Dame Elizabeth Taylor or Ms Fredriksson, but as I understand state-of-the-art medical procedures they would normally both have had one or the other of the many gadolinium compounds developed for this purpose.[184] Taylor's tumour was benign, and she lived for another 14 years, dying at the age of 79 in 2011, but, as far as I understand, the brain surgery following the MRI scan on Fredriksson really saved her life.

Some of you may recognize the elements of the lanthanoid series (or lanthanides—chemists have problems agreeing over this term, although the former is the IUPAC-approved name) where we find gadolinium as one of the *rare earth elements* (REEs). This term suggests that we may shortly find ourselves in trouble manufacturing these compounds, but 'rare' in this case should better be understood as 'exceptional', as even the least abundant of these elements, lutetium and thulium, are almost 200 times more common than gold. The 'crisis' of the REEs in 2010–11, including a brief Chinese exports freeze, had more to do with the closing down of mines and the reluctance to invest in prospecting and new mines in the rest of the world—because we do find these elements all over the globe, not just in China.[185, 186]

Because serious mining has been around ever since Snow White and the Seven Dwarfs, or even longer, we take it for granted, associating it with dynamite, heavy drills, and brave men with head-lamp-crowned helmets. While this is true, and I would not like to diminish the heavy and often dangerous work of miners all over the world, the more complex technology will often be found above ground. Not only can a mine be described as a hole in the ground with a chemical plant on top, but for every element we need to extract—and depending on the exact chemical composition of the compounds we take up from the Earth's interior—the plant needs to be specifically designed. Added to this are the complex environmental and legislative problems almost always associated with the start-up of a new mine. Thus rare earth production is not begun in one a fell swoop. For example, from the start of serious prospecting in 2009, in an already existent but closed exploratory mine in the vicinity of the sleepy Swedish provincial town of Gränna, famous for its peppermint candy canes,* it will take until at least 2016 until the proposed mine is in production and yielding, among other REEs and zirconium, 100–160 tonnes of gadolinium concentrate annually.[187]

---

* In certain conditions and shapes these candies are referred to as 'rocks' in the UK. Oddly in Swedish they are known as 'polkagrisar', the last part of the word meaning pig ('gris'), which, among other things, is a crude casting of metal suitable for further processing, for example pig iron.

21

Of Pea-Soup, Dangers of Coffee in the Morning, and the Test of Mr Marsh

Chapter 21, in which a murderer is acquitted and, as a result, many others are brought to justice. We also learn about reductions and oxidations, and a chameleon group of elements known as the metalloids.

Friday the thirteenth is supposed to be the unlucky day, but for ex-King Eric XIV of Sweden it must have been a Thursday, because Thursdays are pea-soup days, at least in Sweden and Finland. It may only be a persistent myth that the arsenic trioxide (As_2O_3), which probably killed him, was put in his pea-soup on the order of his half-brother John III. That is, the pea-soup bit may be a myth, not that his brother John was the instigator. He had already held his schizophrenic[188] older brother Eric prisoner for nine years, and a number of incriminating documents have been preserved.[189]

Eric died in 1577, and his fate mirrors that of Mary Stuart, who was sentenced to a more conventional execution by her cousin*

---

* Henry VII was Elizabeth's grandfather and Mary's great-grandfather.

Elizabeth I ten years later. Oddly enough Eric had, with a certain hubris one must say, tried to negotiate marriages with both these distinguished ladies, and was only a few days from sailing off to meet Elizabeth in person when his father Gustav Wasa died in 1560 distracting him with other matters for a while.[190]

In 1577 there was no good way to analyse arsenic and establish murder by poison, but in 1829 the situation was different. When John Bodle was tried for having murdered his grandfather, octogenarian George Bodle, on his farm in Plumstead near Woolwich, the prosecution could provide an expert witness, James Marsh, inventor and (among other things) assistant to Michael Faraday. Marsh, through the foresight of the local police, who were already suspicious, and had preserved both the last coffee George Bodle had drunk and his stomach contents, analysed both for arsenic.[191] This he did by adding hydrogen sulphide (H_2S), a foul-smelling, flammable, and poisonous gas that used to haunt undergraduate chemistry labs when I was young.

Dangerous as it may be, rather elementary precautions make it safe to handle H_2S even for first-year students, and it was used in much the same way as by Marsh—for hunting down metal ions. Arsenic(III)—its oxidation state in As_2O_3—is big and soft, situated as it is a fair bit down in the Periodic Table, and, because softies like softies, the H_2S will release its S^{2-} ions that will then rapidly combine with As^{3+} to give the yellow and very insoluble sulphide, As_2S_3. It does much the same with other metal ions, especially the bigger and softer ones, often forming coloured, insoluble solid phases that we call precipitates when they appear in a previously clear solution. Some are bright red, some black, some white, and others yellow. Thus, adding H_2S is a simple visual test to prove the presence of various metal ions.

This was a clear-cut case one would have thought, as Bodle junior had also twice been into the local pharmacy to buy arsenic, commonly available as rat poison, and the maid had testified on his unusual willingness to help fetch water for his grandfather's coffee that particular morning. The problem for the prosecution was, however, that the As_2S_3 had turned old awaiting the trial, and had lost its bright yellow colour. The jury consequently did not accept the technical evidence, and a murderer walked free, inheriting his grandfather's probably rather substantial farm.[192]

This left James Marsh very frustrated, and he set out to invent a better test which relied on a curious property of arsenic: its metallic character. Arsenic is situated below nitrogen and phosphoros, the first a gas in its natural pure state, the other forming distinct P_4 molecules, but as we go down this part of the Periodic Table known as the 'main group elements' the tendency to be metallic increases. Carbon is no metal, but lead is, for example. Arsenic is a typical so-called *metalloid*, or half-metal, and just like tin it can appear in several allotropes, or crystal forms (see Chapter 17), the most stable at room temperature being the grey arsenic that has a metallic appearance. The cue for the new test Marsh devised, which was perfected during the nineteenth century, was that it produces a shiny, mirror-like metal surface on the inside of a test tube, easy to preserve, more distinctive and indeed more sensitive than relying on a yellow powder.[193]

The Marsh test has several more steps, but it is not a difficult procedure to follow. Agatha Christie, as mentioned earlier, learnt it as a novice pharmacy student during World War I, and that man of many talents, Lord Peter Wimsey's valet Mervyn Bunter, does the same in Dorothy Sayers' 1930 novel, *Strong Poison*.[194] More surprisingly is perhaps that it turns up in celebrated children's author Astrid Lindgren's work. Before turning to writing she was the secretary of one 'Revolver-Harry' Söderman, a remarkable policeman with a PhD, who used his position to secretly train a substantial armed force of Norwegian 'policemen' at a special Swedish 'Health Spa' during the German occupation of Norway. He was also one of the founders of modern forensic science *and* of INTERPOL. No doubt Lindgren learned a trick or two from him during World War II.[195]

A few years after writing Pippi Longstocking and, eventually, walking into the history of children's literature, she published a series of three crime novels for adolescents and older children, apparently sparking something of a 'CSI-fever' in 1950s Sweden. In one of these, a young teenager used his knowledge of chemistry to carry out the Marsh test on a piece of chocolate. His name was Kalle Blomkvist, a name that also appears in the Millennium Trilogy* by Stieg Larsson, and Mikael Blomkvist is portrayed by no less than Daniel Craig in the 2010 movie *The Girl with the Dragon Tattoo*.[196, 197]

In fact, the Marsh procedure is so simple that I will give it to you in detail here: the only recipe you will find in this book.

Hydrogen gas (H_2) is passed through the sample. In this compound hydrogen has the oxidation state zero, but it is itching to give away its two electrons (one per H) and become an H^+, either as an ion or in a compound. Arsenic in As_2O_3 has the oxidation state +III and will pick up three electrons from three hydrogen molecules and then combine with the resulting H^+ to give another gas, arsine (AsH_3),† and water. In the words of a chemist the As^{3+}

---

* That the female protagonists in the two trilogies have the similar names Lisander and Salander is probably not a coincidence either.

† AsH_3 is analogous to ammonia, NH_3, but much more poisonous.

is reduced—that is, the oxidation state is lowered (from +III to −III), and hydrogen is oxidized (from 0 to +I). Here is the redox reaction:

$$6H_2 \text{ (g)} + As_2O_3 \text{ (s)} \rightarrow 2\,AsH_3 \text{ (g)} + 3H_2O \text{ (l)}$$

The gas is now liberated from the sample and can travel up a test tube with the surplus H_2 gas. When heated with a flame higher up the tube the arsine gas will decompose to give metallic arsenic and hydrogen gas again:

$$2AsH_3 \text{ (g)} + \text{heat} \rightarrow 2As\,(\text{metal}) + 3H_2\,(g)$$

A note of caution: Agatha Christie blew up a coffee-making machine while attempting the test;[117] as we saw in Chapter 3 hydrogen gas can be explosive.

The metal is initially present as a gas, but will rapidly hit the cooler glass surface and condense, just like water vapour, forming a nice metallic 'mirror' on the inside of the tube.

The huge interest in this reaction, attracting the attention of nineteenth-century chemical heavyweights such as Jöns Jacob Berzelius in Sweden and Justus Liebig in Germany, was not only because of its potential for catching criminals. Arsenic was a problem for two reasons. It was commonly available to anyone as a convenient rat poison and could thus by accident turn up almost anywhere (as the rats did), and more importantly, the great attraction between the two soft ions As^{3+} and S^{2-} meant that iron, often made from sulphide-containing ores, would frequently have low levels of arsenic present. Through various processes these arsenic atoms would be transferred to other chemicals, notably sulphuric acid that then, as now, was the very basis on which many compounds were made. Thus a reliable arsenic test was important in many areas, as it still is today.

Only a few years ago a huge arsenic problem was uncovered in Bangladesh, not because of industrial contamination, but because new bore holes for fresh water to replace unhealthy surface water turned up water with a naturally high arsenic content due to the chemical composition of the minerals found underground.[198,199]

John III, who poisoned his brother with arsenic in the most Machiavellian way, was not remotely as successful as his English counterpart Elizabeth I. His son Sigismund, already elected king of the very Catholic Polish Republic, succeeded him in 1592. Sigismund's uncle Charles, a protestant and younger brother of Eric and John, ousted him in 1599. Only then could Sweden really escape from these medieval power transitions.

22

To Take Back the Future

In this chapter there is talk about a very expensive metal and how to use it to make a costly drug cheaper. It will also tell you how to save the world and about a transport problem in the brain.

There are different ways to be propelled into stardom. In 1953 Audrey Hepburn used a scooter in the William Wyler film *Roman Holiday*. The unsteady ride ends at a police station, and with Hepburn earning an Academy Award for best actress. A rather different approach was taken by Michael Douglas and friends in *Romancing the Stone* (Robert Zemeckis 1984, also produced by Douglas), where Douglas and Kathleen Turner are chased throughout most of the film by Danny DeVito in a white Renault 4L.

These more modest modes of transport were not quite the style of Michael J. Fox in Zemeckis' next movie *Back to the Future*—Fox's vehicle to international fame is a plutonium-powered DeLorean sports car. While Piaggio (the makers of the Vespa used by Hepburn) and Renault are large companies that still exist, the DeLorean Motor Company was already bankrupt in 1982, too early to profit from the success of the movie—a worldwide block-buster sensation in 1985. But even if you could find a used DeLorean DMC-12, the only model ever built by the company, don't expect it to take you back to the 1950s even if you fuel it up with plutonium.

In the movie, Fox's character Marty McFly gets caught up in a time paradox and literally needs to save his own future. In real life, six years later, at the age of 29, Fox was diagnosed with Parkinson's disease, beginning a very real life fight to take back his own future by battling the disease at all levels.

Parkinson's disease is what is known as degenerative neurological disorder. It is chronic, and there is at present no cure, but treatment to combat the symptoms exists. It was first described in detail by English physician James Parkinson, and named after him by the influential Jean-Martin Charcot whom we met briefly in Chapter 12. The classical symptoms are tremors, rigidity, slowness of movements, and balance problems. The problems for doctors, and consequently for their patients, is that there is no simple chemical or biochemical test for Parkinson's disease, sometimes making the diagnosis a complicated affair.

What we do know, however, is that the problems stem from a loss of nerve cells in the brain, and a decrease in production of a signal transmitter molecule called dopamine (shown in Figure 46).

One intuitive way to treat the symptoms would be to simply supply the missing dopamine molecule to the brain. This looks simple on paper, but turns out to be trickier in reality due to two completely unrelated chemical problems.

The first obstacle is the blood–brain barrier that essentially keeps a tight control of which substances are let into the brain's chemical system. If we inject dopamine directly into the bloodstream it will

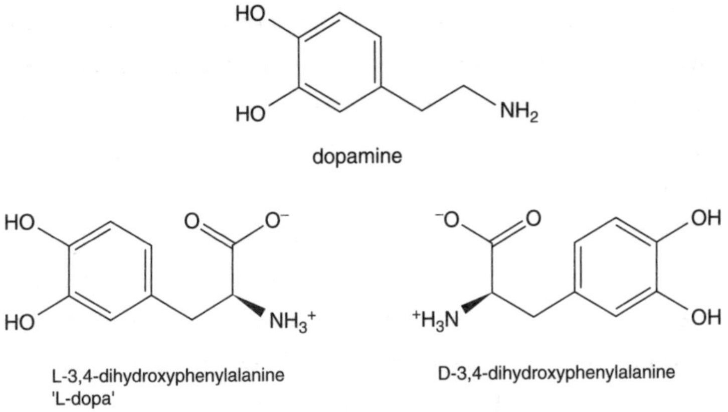

dopamine

L-3,4-dihydroxyphenylalanine
'L-dopa'

D-3,4-dihydroxyphenylalanine

FIGURE 46 Two essential brain molecules and one without any function. The D-3,4-dihydroxyphenylalanine is inactive and the mirror image of L-dopa. The bold black lines mean these bonds are pointing out of the plane of the paper.

bounce off this barrier because it is too much of a water-loving molecule, having what are know as 'polar groups': $-NH_2$ (the amine functionality) and the two $-OH$ (the alcohol groups). What we can do is to supply the brain with the molecule it uses itself to produce dopamine, because this function is still intact.

This molecule is called L-3,4-dihydroxyphenylalanine, or L-dopa for short.[200] If you look at the picture and remember what we have just said, this molecule actually looks like it would have a much lower chance of crossing the blood–brain barrier, as it has not only the alcohol groups but also a charged amino acid end, and anything that is charged will be water-loving. The way it works is that L-dopa will hitch a ride with a protein that specifically recognizes and hides this charged end: to avoid water it will get into a cab to take it across the border.

The next problem is left-hand traffic. When the molecules hitch a ride with the protein cab, it is left-hand traffic only, and the

FIGURE 47 L-dopa and D-dopa, left- and right-handed mirror molecules. Only the leftie is any good against Parkinson's disease.

L-dopa molecules fit perfectly well in the seat to the left of the driver. Which is fine, as long as we have only L-dopa in our drug. L-dopa is also known as levodopa, where the levo stands for 'L' and comes from the Latin *laevus* for left side, as a solution of this compound will turn polarized light anti-clockwise. When we make this chemical in the lab, the trouble is that we would normally get a mixture of the L-dopa and its mirror image D-dopa shown in Figure 47.

Imagine that the big flat part of the L-dopa molecule protruding on its left side is a huge plaster on your left leg and foot, and you need a taxi to take you across the blood–brain frontier. When

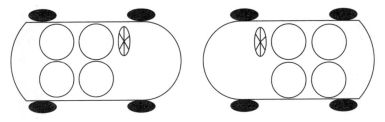

FIGURE 48 Cars for driving on the left side and on the right side of the road are also mirror images—chemists would call them enantiomers.

you sit in the front of the cab, usually there is more space towards the door and it is a bit more cramped towards the driving seat, and consequently you'd be more comfortable in a car with the steering wheel to the right, driving on the left side of the road.

For the real D-dopa molecules this effect is even more pronounced: they will not even get into the car, and even if they did the enzyme that chops off the carboxylic acid part in L-dopa and transforms it into the needed dopamine would not recognize the D-dopa.

So, obviously we should just make L-dopa and give it to patients with Parkinson's disease. However, as it turns out this is easier said than done. These types of molecules that have mirror images that are not identical, just like a right and a left hand, are called optically active (or more correctly chiral),* and could for a long time only be obtained from biological sources. Whenever the chemist in the lab tried his or her hands at it, both the left-handed and the right-handed molecules would turn up in the round bottom flask, a fact cleverly used as a plot device by Dorothy Sayers and Robert Eustace in their novel *The Documents in the Case* published in 1930.[201]

---

* From the Greek word for hand, *kheir*.

But we have come a long way since the 1930s, and today in the lab chemists routinely make either mirror image (or enantiomers, as we prefer to call them) of a large number of molecules. It can still be challenging and costly, especially for large-scale production, so many of my organic chemistry friends pursue research careers trying to find new and better methods of doing this.

Luckily, for the sufferers of Parkinson's disease, the L-dopa problem was solved as early as in the eventful year of 1968 by William Knowles, an industrial chemist at Monsanto, a US-based chemical company. It was then already clear that this molecule would be in high demand, much higher than any natural sources could provide, and industrial production began in 1974. This was 14 years after L-dopa was first suggested as a treatment for Parkinson's disease by Arvid Carlsson from the University of Gothenburg.

I will not go into details about any of these two gentlemen's work—it is adequately described elsewhere, notably on the Nobel Foundation's website as both Carlsson (Physiology or Medicine 2000) and Knowles (Chemistry 2001) became Nobel Laureates for their research on L-dopa. However, a central point of Monsanto's commercial L-dopa process is catalysis using the precious metal rhodium, and catalysis is such a central concept in modern chemistry that it merits a further look.

We are by now at least vaguely familiar with the concept of a catalyst—a substance that will make a chemical reaction go much faster without itself being consumed in the process—but perhaps not with the fundamental importance it will have for us in the future. If a chemical reaction does not go fast enough, what do we do in the absence of a suitable catalyst? We kick it, most often with heat, and if we do this 24/7 it will cost us dearly in energy. Another problem is what we call yield and by-products. A better catalyst will convert more of our starting materials to useful products and give less waste. Of equal importance, there will be fewer problems in separating the good stuff from the unwanted molecules—a task with major energy demands in most commercial chemical processes.

What Knowles and his team did was to design a rhodium-containing molecule with one side of the small metal ion shielded with an organic molecule that uses phosphorus to attach itself to rhodium.[202] This organic molecule already has a 'handedness', and while the reaction is carried out on the rhodium atom this handedness determined which of the two possible dopa molecules was

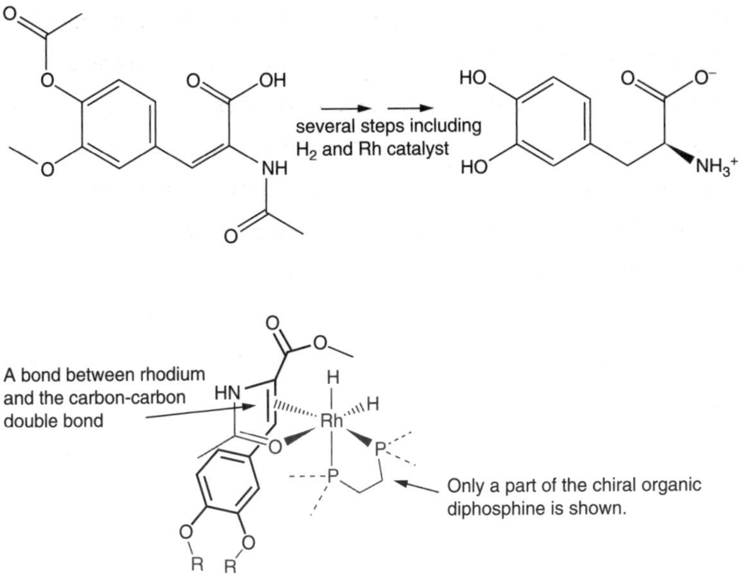

FIGURE 49 Top: schematic reaction diagram showing an achiral starting material transformed to an enantiomerically (optically) pure product. This reaction is carried out in several steps, but the chirality is created by the rhodium catalyst. Bottom: schematic drawing of the chiral rhodium catalyst in action. Bold lines show the atoms that will become the L-dopa molecule.[203]

produced. After reaction with hydrogen gas and the starting material, one molecule of L-dopa will detach itself from the rhodium atom, enabling the whole assembly of rhodium, plus its attached phosphorous organic molecule, to go back to its starting state and the whole process is ready to begin anew.

A catalyst only has one important restriction: it is not a philosopher's stone, and catalysis cannot break the laws of thermodynamics. It is impossible to find a catalyst that would convert water and carbon dioxide into fuel because, in thermodynamic language, it is an uphill reaction. The only reactions we can do are those that are downhill in terms of Gibbs Free Energy (that we encountered when discussing Napoleon's buttons), which is not quite the same thing as saying that all spontaneous reactions produce heat.

Thermodynamics, especially chemical thermodynamics, is a fascinating subject, but many students find it difficult. I think part of the problem is that it starts off by stating the obvious in great mathematical detail, which makes everyone fall asleep, and when they wake up again the lecturer is well into partial derivatives with symbols such as S, G, and μ, and it is very difficult to catch up. This is like having someone explain how cricket works, falling asleep while being told that the batsman has to hit the ball, and waking up later when the game is already underway.

To me, thermodynamics is common sense with mathematical glasses on. Suddenly you can see things in a much sharper way, explore relations otherwise hidden, and arrive at conclusions you were previously blind to. For example, starting with known amounts of molecules A and B, say carbon dioxide and water, you can calculate how much C and D, say octane (a hydrocarbon molecule in petrol) and oxygen, you can make at any given temperature: zero as it turns out, because for Gibbs Free Energy the journey is uphill all the time. Too bad for us—no such shortcuts to solve global warming.

So catalysis cannot change thermodynamics, but it is to chemical reactions what civil engineering is to the Alps: you do not need to cross the mountain passes to get to the Mediterranean, you can pass through the Simplon Tunnel. Catalysts provide nice downhill shortcuts that bypass what chemists call kinetic barriers or activation energies (illustrated in Figure 50). These are much like our own activation barriers: it is nice to lie on the sofa but nicer to fall asleep in bed, although it takes a certain effort to get there.

Catalysis is not only something that happens in industry and in a catalytic converter in a car. It happens with every breath you take. Organic molecules such as carbohydrates, fats, and sugars supply the energy we need in our bodies. When these are digested, energy is converted into usable forms that drive our biochemical processes, and carbon dioxide and water are produced as waste. The carbon dioxide is in the form of bicarbonate ions (HCO_3^-),

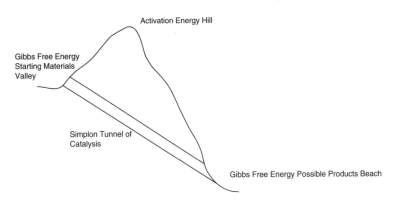

FIGURE 50 Calculation of Gibbs Free Energy tells us if it is downhill to the beach but not how to get there. Travelling through the Tunnel of Catalysis will be the best solution.

and these need to take up an H^+ ion and split into water and gaseous CO_2 before we can breathe out the carbon dioxide through our lungs. You can watch this reaction happen if you add lemon juice (being acidic it will provide the H^+ ions) to carbonated water, and although it looks like the reaction happens instantaneously—as we see the bubbles of CO_2 form immediately—in biochemical terms it much too slow to enable our breathing to function. In our bodies we have a zinc-containing enzyme called carbonic anhydrase that catalyses this reaction, and it is one of the most efficient of all known enzymes.

In general terms, an enzyme is a protein that works as a catalyst, and in the brain there is another one that will take on the L-dopa molecules and convert them to the active neurotransmitter dopamine, and that is how sufferers from Parkinson's disease get some of their relief.

The making of dopamine from L-dopa may be a downhill reaction in terms of Gibbs Free Energy, but making the L-dopa molecules in the body probably is not. How does our biochemistry deal with this? Is it a special feature of being alive? No, what the body does is to couple reactions chemically, somewhat like wagons without their own motors that are linked to an engine to form a train.

In Figure 51, the rightmost little train can effortlessly run by itself down to the *Beach of Low Energy Products*. However, the excess energy will just be transformed to heat and vanish out into the universe, and there is no way we can use that energy to move the leftmost little train unless we couple these reactions chemically through some common reagent or intermediate product. This gives us the tag-cable that will move this train from the *Valley of Stable Starting Materials* to the *High Plateau of Desired Molecules*.

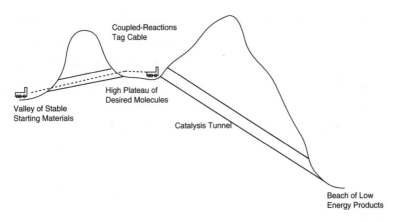

FIGURE 51 By letting the right train pull the left train, we can get the left train to the High Plateau of Desired Molecules while the right train descends to the Beach of Low Energy Products.

Chemists now know a lot about the principles of catalysis and how to use metals, organic molecules, and enzymes to achieve what we need, but we also know that there is much left to do. Perhaps the great work that is being done by the Michael J. Fox Foundation for Parkinson's Research*[204] will be helped by some new innovative catalysed process. Perhaps the solution to Parkinson's disease lies elsewhere in stem-cell research, or more biological approaches, than in small-molecule drugs. There is hope, so long as we can keep our precious high-tech society running by supplying enough energy.

This global energy question also has catalysis as an answer. Enormous hidden supplies of coal cannot save us in the long run, not even with efficient carbon dioxide removal. Fission nuclear energy may be relevant in the short term, but few see it

---

* Michael J. Fox has an honorary doctorate from the Karolinska Institute in Stockholm.

as a permanent solution, and fusion nuclear energy is the joker in the pack we cannot trust. What we know is possible, on the other hand, is the reaction between the energy of the Sun's rays in the form of high-energy photons, and water, to give hydrogen and oxygen gas, H_2 and O_2. It is a simple matter of calculating the Gibbs Free Energy of the process:

$$photon + 2H_2O \rightarrow 2H_2 + O_2$$

For photons found in visible and UV light, we find this process to be downhill indeed. We could laugh all the way to the beach, if we just found the right catalyst. 'Artificial photosynthesis' is the catchy name we have for this, and it has been a kind of Holy Grail for chemists for a long time. In the classic comedy *You Can't Take it with You* from 1938, directed by the chemical engineer Frank Capra,* James Stewart confesses to Jean Arthur that understanding photosynthesis is what he really wants do to, instead of just amassing money like the rest of his family.

Capra does not give us a solution however. The scene is there to prove the noble nature of Stewart's character, Tony Kirby, to his fiancée to be, Alice Sycamore, but technological thriller writers Clive and Dirk Cussler boldly do just that in the novel *Arctic Drift* from 2004.[206] Not inclined to spare the fireworks with their plot devices, they throw in the Franklin expedition's lost ships *The Terror* and *The Erebus* (see Chapter 18) for effect, and base the solution to the world's energy problems on ruthenium catalysis, with an oil and gas magnate—who would be a worthy adversary to James Bond—as the evil genius.

---

* He graduated from the Throop College of Technology, now the California Institute of Technology, in 1918.[205]

The epilogue statement that 'The Kitimat Photosynthesis Station will safely and efficiently convert carbon dioxide to water and hydrogen without any risk to the environment' makes clear that more imagination than science has been put into this novel. Where did the carbons go and where does the hydrogen come from if you start with CO_2 and end with H_2O and H_2? However, the authors use the transition metal ruthenium, left of rhodium in the Periodic Table, as the catalyst. This is not far off the mark, as ruthenium is actually a favourite with chemists working to create systems that could harvest the sun's rays and convert them into usable energy. Green leaves do not use ruthenium though—for one thing, there would not be enough of it. Instead, the ions of the metals magnesium, iron, copper, and manganese are used in the very complex enzyme machinery scientists call *Photosystem I* and *Photosystem II*.

Critically acclaimed British writer Ian McEwan is less clear on what his antihero Michael Beard uses as the catalyst in the 2010 novel *Solar*.[207] This abominable physics Nobel Laureate without a conscience and very little scientific creativity left, steals an invention from a younger colleague who conveniently (for Beard as it turns out) drops dead in his living room. In the end we are left in the dark as to whether the invention will really be able to catalyse the reaction of photons plus water to oxygen and hydrogen gas, but a fair bit of accurate physics has gone into this book. McEwan tells us himself: 'I think one of the reasons I find a lot of novels boring is that they're only about the emotions; they don't have enough muscular intelligence. I like novels that have got both. A good number of novels are just so timid, intellectually.'[208]

Both rhodium and ruthenium are very expensive metals, but as catalysts they can be used again and again—that is the point. For future energy systems, catalysis will be important. For curing Parkinson's Disease we don't know, but right now catalysis provides great relief to many sufferers.

For these and many other similar reasons it is now time to end the book, put on the lab coat, and get on with teaching and research, for never before has the world needed chemistry and professionals educated in chemistry as it does today.

ACKNOWLEDGMENTS

My foremost thanks goes to my family, Nina Kann, Agnes Öhrström Kann, and Rebecka Öhrström Kann, who have had to put up with this project for too long a time.

Thanks also to Latha Menton, Emma Ma, and the publishing team at Oxford University Press, Bernadette Plissart at the Swedish-Finnish-Scottish writers and artists colony in Grez-sur-Loing outside Fontainbleau, and Peter Cottino and his staff at the Axel Munthe Villa San Michele Foundation on Capri.

Many people have been reading selected parts of the manuscript, and I thank you all for correcting my mistakes and making valuable comments. However, you should in no way be associated with the errors and omissions that undoubtedly remain, these are entirely of my own making. My sincere thanks to: Nina Kann, Claes-Rune Öhrström, Göran Svensson, Christian Ekberg, Neil Champness, Deborah Kays, Anna Börje, Jerker Mårtensson, Vratislav Langer, Per Lincoln, Per Cullhed, Birthe Sjöberg, Torbjörn Granlund, Gunnar Westman, Göran Petersson, Hans Nissbrandt, Anna Said, Alireza Movahedi, Susan Bourne, Linda Kann, Jan Reedijk, Vadim Kessler, Peter Stilbs, Lars Bentell, Britt Marie Hartvig, Johanna Nganunu Macharia, Claes Niklasson, Isabelle Michaud-Soret, Marie Brigantini, Petter Djerf, Antii Laurila, Elina Laurila, Berndt Björlenius.

For assisting with enquires in various ways I thank the late Theodore Rockwell, Bjarne Bekker, Frank Delaney, Mary Greene, Daniel Rabinovitch, Eugenijus Butkus, Rimantas Jankauskas, Adam Zamoyski, John Williams, Yngve Axelsson, Richard Van Treuren, Wilco Keur, Anders Edling, The Lithuanian National Museum in Vilnius, The Manuscript Collections at Uppsala University Library, Chalmers Library, and Jernkontoret in Stockholm.

For financial assistance I want to thank the Hasselblad Foundation for a writer's stipend to Hôtel Chevillon in 2008, the Axel Munthe Villa San Michele foundation for subsidizing my stay at Villa San Michele on Capri 2011, and the Royal Society of Chemistry for an International Author Grant in 2012. Generous support from the Chalmers Foundation during 2012 is gratefully acknowledged.

BIBLIOGRAPHY

For general references to the chemical industry and chemical engineering I have used the multi-volume encyclopaedias that are standard tools for the chemical engineer and contain substantial articles on anything chemical, from nuclear energy to fragrances:

Kirk-Othmer Encyclopedia of Chemical Technology. (John Wiley & Sons, New York 1999–2012).

Ullmann's Encyclopedia of Industrial Chemistry. (Wiley-VCH Verlag Gmbh & Co. Weinheim, 1999–2013).

For general references to chemistry:

F. A. Cotton, G. Wilkinson, Advanced Inorganic Chemistry. (Wiley, New York, 1989).

N. N. Greenwood, A. Earnshaw, Chemistry of the Elements. (Pergamon Press, Oxford, 1997).

For the history of chemistry and the chemical industry I have on my shelf:

A. J. Ihde, The Development of Modern Chemistry. (Dover Publications, New York, 1984).

Other more general references:

J. Emsley, Nature's Building Blocks: An A—Z Guide to the Elements. (Oxford University Press, Oxford, 2003).

J. Emsley, The Elements of Murder. (Oxford University Press, Oxford, 2005).

For general information on movies and actors I have mostly used:

L. Maltin, Leonard Maltin's 2010 Movie Guide. (Plume, New York, 2009).

D. Thomson, The New Biographical Dictionary of Film. (Knopf, New York, 2003).

REFERENCES

1. *IUPAC Periodic Table of the Elements*. (The International Union of Pure and Applied Chemistry, 2012). <http://www.iupac.org/fileadmin/user_upload/news/IUPAC_Periodic_Table-1Jun12.pdf> accessed 13 June 2013.
2. E. R. Scerri, *The Periodic Table: Its Story and Its Significance*. (Oxford University Press, Oxford, 2006).
3. P. Ball, 'There is No Hidden Understanding to be Teased out by "Improving" the Periodic Table, Argues Philip Ball. But Eric Scerri Begs to Differ,' *Chemistry World*, September 2010.
4. D. Brown, *The Da Vinci Code*. (Doubleday, New York, 2003).
5. D. Adams, *The Hitchhiker's Guide to the Galaxy*. (Pan Books, London, 1979).
6. G. Mwakikagile, *Botswana Since Independence*. (New Africa Press, Pretoria, 2009).
7. N. Parsons, W. Henderson, T. Tlou, *Seretse Khama*. (Macmillan, Gaborone, 1995).
8. W. Mbanga, T. Mbanga, *Seretse & Ruth*. (Tafelberg, Cape Town, 2005).
9. M. Dutfield, *A Marriage of Inconvenience*. (HarperCollins, New York, 1990).
10. R. J. Duffy, 'History of Nuclear Power,' in *Encyclopedia of Energy*. (Elsevier, Amsterdam, 2004), vol. 4.
11. ''49 Uranium Rush,' *Popular Mechanics* 91, no. 2, February 1949.
12. R. P. H. Willis, 'The Uranium Story—An Update,' *The Journal of The South African Institute of Mining and Metallurgy* 106, 601–9 (2006).
13. Z. Masiza, 'A Chronology of South Africa's Nuclear Program,' *The Nonproliferation Review* Fall, 35–55 (1993).
14. R. Hyam, 'The Political Consequences of Seretse Khama: Britain, the Bangwato and South Africa, 1948–1952,' *The Historical Journal* 29, 921 (1986).
15. R. Hyam, P. Henshaw, 'Prime Minister's Office Papers, PRO, PREM 8/1308, minutes by Attlee, 21 December 1949 and 22 January 1950,' cited

in *The Lion and the Springbok: Britain and South Africa Since the Boer War*. (Cambridge University Press, Cambridge, 2003).

16. A. McCall Smith, *The Number One Ladies Detective Agency*. (Polygon, Edinburgh, 1999).

17. R. Hyam, P. Henshaw, *The Lion and the Springbok: Britain and South Africa Since the Boer War*. (Cambridge University Press, Cambridge, 2003).

18. D. Campbell, 'Alabama Votes on Removing its Ban on Mixed Marriages Special Report: The US Elections', *The Guardian*, 3 November 2000.

19. Obituary, 'Henry S. Lowenhaupt, 87; CIA Trailblazer', *The Washington Post*, 14 March 2006.

20. H. S. Lowenhaupt, 'Chasing Bitterfeld Calcium', *Studies in Intelligence* 17, no. 1 (1996 (released as sanitized, original published in 1973)).

21. M. S. Goodman, *Spying on the Nuclear Bear: Anglo–American Intelligence and the Soviet Bomb*. (Stanford University Press, Stanford, 2007).

22. C. Campbell, *A Questing Life*. (iUniverse, Bloomington, 2006).

23. B. Brinck, *Stockholms-Tidningen*. Swedish reporter onboard the Hindenburg 1937, various articles sent home to Stockhom during the last voyage and cited in the 2011 *Sveriges Radio* documentary 'Ombord på Hindenburg', 30 January 2010.

24. P. Russell, *Faces of the Hindenburg*. (2013). <http://facesofthehindenburg.blogspot.com> accessed 10 June 2013.

25. *The Hydrogen and Fuel Cells Program*. (U.S. Department of Energy, 2013). <http://www.hydrogen.energy.gov/> accessed 10 June 2013.

26. Personal Communication. Richard van Treuren using data from: H. v. Schiller, *Zeppelin, Wegbereiter des Weltluftverkehrs*. (Kirschbaum Verlag, Bad Godesberg, 1966).

27. Hindenburg footage from aboard the last flight, *YouTube*, <http://www.youtube.com/watch?v=orSzc8JxBFg> accessed 10 June 2013.

28. W. N. Medlicott, 'The Economic Blockade', in *History of the Second World War, United Kingdom Civil Series*, W. K. Hancock, ed. (Longmans, Green and Co., London, 1952).

29. J. T. Woolley, G. Peters, *Herbert Hoover, Statement About the Export of Helium at the President's News Conference, October 10, 1930, The American Presidency Project, Santa Barbara, CA: University of California* (1930). <http://www.presidency.ucsb.edu/ws/?pid=22382> accessed 10 June 2013.

30. N. Shute, *Slide Rule*. (Heinemann, London, 1954).

31. G. J. White, J. R. Maze, *Harold Ickes of the New Deal: His Private Life and Public Career*. (Harvard University Press, Harvard, 1985).

32. J. T. Woolley, G. Peters, *Recommendation on a Policy for Helium Export, May 25, 1937, The American Presidency Project*, Santa Barbara, CA: University of California (1937). <http://www.presidency.ucsb.edu/ws/?pid=15408> accessed 10 June 2013.

33. 'Helium to Germany,' *Time Magazine*, Monday 17 January 1938.

34. H. L. Ickes, *The Secret Diary of Harold L. Ickes: The Inside Struggle: 1936–1939—Volume 2.* (Simon and Schuster, New York, 1955).

35. P. Berg, 'Background and Biography,' in *R.R. Angerstein's Illustrated Travel Diary, 1753–1755: Industry in England and Wales from a Swedish Perspective*, P. Berg, ed. (Science Museum, NMSI Trading Ltd, Exhibition Road, London, 2001).

36. R. R. Angerstein, *R.R. Angerstein's Illustrated Travel Diary, 1753–1755: Industry in England and Wales from a Swedish Perspective*. P. Berg, ed. (Science Museum, NMSI Trading Ltd, Exhibition Road, London, 2001).

37. S. Rydberg. For more on later espionage efforts against Huntsman, see the travels of Johan Ludvig Robsahm, 1761, pp. 189–91, in *Svenska studieresor till England under frihetstiden.* (Almquist & Wiksells boktryckeri, Uppsala and Stockholm, 1951), pp. 170–87.

38. W. Scott, *The Talisman*. (Archibald Constable and Co., and Hurst, Robinson and Co., Edinburgh, 1825).

39. M. Reibold, P. Paufler, A. A. Levin, W. Kochmann, N. Patzke, D. C. Meyer, 'Materials—Carbon nanotubes in an ancient Damascus sabre,' *Nature* 444, 286 (2006).

40. S. Mader, 'Scott's "Talisman", Damask Salad and Nano-wires—Observations to the Fundamental Natural Scientific Studies of Phantoms,' *Waffen-Und Kostumkunde* 49, 45 (2007).

41. M. Fornander, 'Biografi,' in *Reinhold R. Angersteins resor genom Ungern och Österrike 1750*, M. Fornander, ed. (Jernkontorets bergshistoriska utskott, Jernkontoret, Stockholm, 1992).

42. E. Naumann, 'Reinhold Rüdker Angerstein,' in *Svenskt biografiskt lexikon* (Riksarkivet, Stockholm, 1918), vol. Band 01, p. 792.

43. M. Palmer, Introduction, in *R.R. Angerstein's Illustrated Travel Diary, 1753–1755: Industry in England and Wales from a Swedish Perspective*, P. Berg, ed. (Science Museum, NMSI Trading Ltd, Exhibition Road, London, 2001).

44. S. M. Farrell, Wentworth, Mary Watson-, marchioness of Rockingham (bap. 1735, d. 1804), in *Oxford Dictionary of National Biography*. (Oxford University Press, Oxford, 2004).

45. M. Proust, *Swann's Way: Remembrance of Things Past*. (Courier Dover Publications, Mineola, 2002, original published in 1913).
46. G. Milton, *Nathaniel's Nutmeg*. (Penguin Books, London, 1999).
47. 'International Trade Centre, World Markets in the Spice Trade 2000–2004,' *UNCTAD/WTO* (2005).
48. United Nations Convention on Biodiversity, *The Convention on Biological Diversity*. 120901 <http://www.cbd.int/> accessed 1 September 2012.
49. H. Andersson, *Från dygdiga Dorothea till bildsköne Bengtsson berättelser om brott i Sverige under 400 år*. (Vulkan, Stockholm, 2009).
50. *Mining Area of the Great Copper Mountain in Falun, UNESCO World Heritage site*. (UNESCO, 2001). <http://whc.unesco.org/en/list/1027> accessed 6 April 2013.
51. G. Mwakikagile, *Africa 1960–1970: Chronicle and Analysis*. (New Africa Press, Pretoria, 2009).
52. 'Chilean President Salvador Allende Committed Suicide, Autopsy Confirms,' *The Guardian* (20 July 2011).
53. J. Borger, 'New Inquiry Set Up Into Death of UN Secretary General Dag Hammarskjöld,' *The Guardian*, 18 July 2012.
54. V. Finlay, *Colour: Travels Through the Paintbox*. (Sceptre, London, 2003).
55. M. Polo, L. F. Benedetto, *The Travels of Marco Polo: Translated into English from the Text of L.F. Benedetto*. (Asian Educational Services, New Dehli, 1931).
56. J. Wood, *A Personal Narrative of a Journey to the Source of the River Oxus*. (John Murray, London, 1841).
57. T. Perry, 'Afghan Commander Massoud, Killed on Eve of 9/11 Attacks, is a National Hero Proclaimed September 9 as Massoud Day,' *Los Angeles Times*, 22 September 2010.
58. Obituary, 'Ahmad Shah Massoud,' *The Telegraph*, 17 September 2001.
59. M. Dupee, *Afghanistan's Conflict Minerals: The Crime–State–Insurgent Nexus*. (United States Military Academy, 2012). <http://www.ctc.usma.edu/posts/afghanistans-conflict-minerals-the-crime-state-insurgent-nexus> accessed 11 December 2012.
60. G. Hildebrandt, 'The Discovery of the Diffraction of X-rays in Crystals—A Historical Review,' *Crystal Research and Technology* 28, 747 (1993).
61. I. Hassan, R. C. Peterson, H. D. Grundy, 'The Structure of Lazurite, Ideally $Na_6Ca_2(A_{16}Si_6O_{24})S_2$, a Member of the Sodalite Group,' *Acta Crystallographica* C41, 827 (1985).
62. D. Arieli, D. E. W. Vaughan, D. Goldfarb, 'New Synthesis and Insight into the Structure of Blue Ultramarine Pigments,' *Journal of the American Chemical Society* 126, 5776 (2004).

63. M. E. Fleet, X. Liu, X-ray 'Absorption Spectroscopy of Ultramarine Pigments: A New Analytical Method for the Polysulfide Radical Anion S_3^- chromophore,' *Spectrochimica Acta* Part B 65, 75 (2010).

64. A. M. Smith, *Tears of the Giraffe*. (Polygon Books, Edinburgh, 2000).

65. F. Delaney, *Simple Courage: A True Story of Peril on the Sea*. (Random House, New York 2006).

66. B. Bekker, *Flying Enterprise & Captain Carlsen*. (Also available in English translation as Flying Enterprise & Captain Carlsen). (Bekkers forlag, Skårup, 2011).

67. T. Rockwell, *The Rickover Effect: How One Man Made a Difference*. (Naval Institute Press, Annapolis, 1992).

68. H. G. Rickover, L. D. Geiger, B. Lustman, 'History of the Development of Zirconium Alloys for Use in Nuclear Reactors,' (United States Energy Research and Development Administration, Division of Naval Reactors, Washington, 1975).

69. E. T. Hayes, 'Part VII—Alloys,' in *Zirconium: Its Production and Properties, U.S. Bureau of Mines Bulletin 561*. (1956), p. 93.

70. T. Rockwell, Personal Communication (2012).

71. F. L. S. Bowman, *Statement of Admiral F. L. 'Skip' Bowman, U.S. Navy Director, Naval Nuclear Propulsion Program Before the House Committee on Science 29 October 2003*. (U.S. Navy, 2003). <http://www.navy.mil/navydata/testimony/safety/bowman031029.txt> accessed 13 June 2011.

72. H. Petroski, *The Pencil: A History of Design and Circumstance*. (Knopf, New York, 2006).

73. I. Tyler, *Seathwaite Wad, and the Mines of the Borrowdale Valley*. (Blue Rock Publications, Cumbria, 1995).

74. W. G. Collingwood, *Lake District History*. (Titus Wilson & Son, Printers, 1925).

75. *An Act for the More Effectual Securing Mines of Black Lead from Theft and Robbery, The Statues at Large from the Twentieth Year of Reign of King George the Second to the Thirtieth Year of Reign of King George the Second*. (Mark Basket, London, 1764), vol. 7, p. 415.

76. J. P. Pederson, *International Directory of Company Histories, Volume 73*. (Gale, Farmington Hills, 2005).

77. G. Petrie, *Hand of Glory*. (Pan Books, London, 1979).

78. H. Walpole, *Rouge Herries*. (Macmillan, London, 1932).

79. P. Høeg, *Miss Smilla's Feeling for Snow*. (Vintage, London, 1996 (English ed.)).

80. A. v. Hees, 'Fiction and Reality in Smilla's Sense of Snow,' *European Studies, An Interdisciplinary Series in European Culture, History and Politics* 18, 215 (2002).

81. T. Geller, 'Aluminum: Common Metal, Uncommon Past,' *Chemical Heritage Magazine*, 2007/8.

82. G. I. Kenney, *Dangerous Passage: Issues in the Arctic*. (Natural Heritage Books, Toronto, 2006).

83. A. K. Sørensen, *Denmark-Greenland in the Twentieth Century*. (Museum Tusculanum Press, Copenhagen, 2009).

84. G. Lorentz, S. Hommerberg, *Bernt Balchen—Den flygande vikingen*. (AB Allhems förlag, Malmö, 1945).

85. 'Representative Missions: Heroya,' Army Air Forces Report, 4 January 1944. <http://www.nationalmuseum.af.mil/factsheets/factsheet.asp?id=1704> accessed 13 June 2011.

86. R. W. Bo Widfeldt, *Making for Sweden: The Story of the Allied Airmen Who Took Sanctuary in Neutral Sweden. P. 2 The United States Army Air Force*. (Air Research Publications, Walton on Thames, 1998).

87. L. Ramberg, *Kyoto och fjärilarna*. (Kabusa Böcker, Göteborg, 2007).

88. P. Forster, V. Ramaswamy, P. Artaxo, T. Berntsen, R. Betts, D. W. Fahey, J. Haywood, J. Lean, D. C. Lowe, G. Myhre, J. Nganga, R. Prinn, G. Raga, M. Schulz, R. V. Dorland, 'Changes in Atmospheric Constituents and in Radiative Forcing,' in *Climate Change 2007: The Physical Science Basis. Contribution of Working Group I to the Fourth Assessment Report of the Intergovernmental Panel on Climate Change*, S. Solomon et al., eds. (Cambridge University Press, Cambridge, 2007).

89. K. Secher, O. Johnsen, 'Minerals in Greenland,' in *Geology and Ore*. (Geological Survey of Denmark and Greenland (GEUS), Copenhagen, 2008).

90. *He Did Not Know of the A-bomb Plan: A Story about Joachim Rønneberg, and Other Documents in English and Norwegian on the Rønneberg Family Website*. <http://www.ronneberg.org/> accessed 12 December 2011.

91. N. Thomas, *Foreign Volunteers of the Allied Forces 1939–45*. (Osprey Publishing, Oxford, 1991).

92. R. V. Jones, *Most Secret War*. (Hamish Hamilton, London, 1978).

93. U. Uttersrud, *Leif Tronstad Vitenskapsmann, etterretningsoffiser og militær organisator 1903–1945*. (Teknologihisotria, Engineering education, Oslo and Akershus University College of Applied Sciences). <http://www.iu.hio.no/~ulfu/historie/tronstad/> accessed 11 December 2012.

94. K. Okkenhaug, 'NTH-professoren som snøt Hitler for atombomben,' *Adresseavisen*, 14 March 2008.

95. M. Nordahl, 'Tungtvannsaksjonen som mislyktes,' *forskning.no*, an online newspaper devoted to Norwegian and international research, (2011).

96. L. Sjöstrand, 'Tegnér, Strindberg och Fröding—Diktare under psykiatrins lupp,' *Läkartidningen* 102, 660–2 (2005).

97. J. Owen, 'Stephen Fry: My Battle with Mental Illness, The Comic Actor Talks Openly for the First Time About the Self-loathing Brought About By His Bipolar Disorder,' *The Independent*, 17 September 2006.

98. J. A. Quiroz, T. D. Gould, H. K. Manji, 'Molecular Effects of Lithium,' *Molecular interventions* 4, 259 (2004).

99. L. Hultqvist, *Strindberg: Guldmakaren in Nationalencyklopedin.* (2004). <http://www.ne.se/rep/strindberg-guldmakaren> accessed 25 March 2013.

100. A. Strindberg, *Hemsöborna* (*The People of Hemsö*). (Albert Bonniers förlag, Stockholm, 1887).

101. Editorial, 'Across the great divide,' *Nature Physics* 309 (2009).

102. G. B. Kauffman, 'August Strindberg, Goldmaker,' *Gold Bulletin* 21, 584 (1983, 1988).

103. H. B. d. Saussure, *Voyages dans les Alpes.* (Samuel Fauche, 1786), vol. II.

104. B. S. Hetzel, 'The Nature and Magnitude of the Iodine Deficiency Disorders,' in *Towards the Global Elimination of Brain Damage Due to Iodine Deficiency,* B. S. Hetzel, ed. (Oxford University Press, Oxford, 2004).

105. J. Orgiazzi, S. W. Spaulding, *Milestones in European Thyroidology (MET) Jean-Francois Coindet (1774–1834).* (The European Thyroid Association). <http://www.eurothyroid.com/about/met/coindet.php> accessed 26 March 2013.

106. L. Rosenfeld, 'Discovery and Early Uses of Iodine,' *Journal of Chemical Education* 77, 984 (2000).

107. C. Beckers, *Milestones in European Thyroidology (MET) Introduction.* (The European Thyroid Association). <http://www.eurothyroid.com/about/met/introduction.php> accessed 26 March 2013.

108. G. Droin, 'Endemic Goiter and Cretinism in Alps: Evolution of Science and Treatments, Transformation of the Pathology and its Representations,' *International Journal of Anthropology* 20, 307 (2005).

109. D. G. McNeil Jr, 'In Raising the World's I.Q., the Secret's in the Salt,' *New York Times*, 16 December 2006.

110. *Micronutrients, Macro Impact: The Story of Vitamins and a Hungry World.* (Sight and Life Press c/o Sight and Life/DSM Nutritional Products Ltd, PO Box 2116, 4002 Basel, Switzerland).

111. C. G. Goetz, M. Bonduelle, T. Gelfand, *Charcot: Constructing Neurology.* (Oxford University Press, Oxford, 1995).

112. L. S. Goodman, A. Gilman, *The Pharmacological Basis of Therapeutics: A Textbook of Pharmacology, Toxicology, and Therapeutics for Physicians and Medical Students.* (Macmillan, London, 1970).

113. E. N. Brandt, *Growth Company: Dow Chemical's First Century.* (Michigan State University Press, East Lansing, 1997).

114. A. Christie, *The Mysterious Affair at Styles.* (John Lane, London, 1920).

115. O. Matsson, *En dos stryknin: om gifter och giftmord i litteraturen.* (Bokförlaget Atlantis, Stockholm, 2012).

116. J. H. Robertson, C. A. Beevers, 'The Crystal Structure of Strychnine Hydrogen Bromide,' *Acta Crystallographica* 4, 270 (1951).

117. A. Christie, *Agatha Christie: An Autobiography.* (William Collins Sons & Co Ltd, Glasgow, 1977).

118. E. Verg, G. Plumpe, H. Schultheis, *Meilsteine, 125 Jahre Bayer 1863–1988.* (Informedia Verlags-GmbH, Köln, 1988).

119. *Database of the Imperial War Graves Commission.* (2013). <http://www.cwgc.org/> accessed 12 June 2013.

120. M. Arthur, *Dictionary of Explosives.* (P. Blakiston's Son & Co, Philadelphia, 1920).

121. J. Williams, *From Corn to Cordite.* (John Williams, 2010). <http://corntocordite.weebly.com/> accessed 12 June 2013.

122. D. Lloyd George, *War Memoirs of David Lloyd George.* (Ivor Nicholson & Watson, London, 1933).

123. R. Bud, *The Uses of Life: A History of Biotechnology.* (Cambridge University Press, Cambridge, 1994).

124. T. Keshav, *Biotechnology.* (John Wiley & Sons (Asia) Pte Ltd, Singapore, 1990).

125. G. B. Kauffman, I. Mayo, 'Chaim Weizmann (1874–1952): Chemist, Biotechnologist, and Statesman, the Fateful Interweaving of Political Conviction and Scientific Talent,' *Journal of Chemical Education* 71, 209 (1994).

126. G. B. Shaw, *Arthur and the Acetone: Complete Plays Vol. III.* (Dodd, Mead & Company, New York, 1962).

127. C. Weizmann, *Trial and Error.* (Harper, New York, 1949).

128. L. Burr, *British Battlecruisers 1914–1918.* (Osprey Publishing, Oxford, 2006).

129. D. K. Brown, *The Grand Fleet: Warship Design and Development, 1906–1922* (Naval Institute Press, 2010, reprint of original published by Chatham, London, 1999).

130. *Loftus Jones VC: A Biography.* (Royal Naval Museum Library, Portsmouth, 2005).

131. O. Langlet, 'Salpetersjuderiet och salpetersjudarna,' *Från Borås och de Sju häraderna* 29 (1975).

132. S. Arrhenius, *Kemien och det moderna livet*. (H. Gebers, Stockholm, 1919).

133. *Tidskriften KGF-nytt, Kronobergs Genealogiska Förening* (2006).

134. 'Upphittad dödskalle,' *Agunnaryds Allehanda*, 2010.

135. D. Cressy, *Saltpeter: The Mother of Gunpowder*. (Oxford University Press, Oxford, 2013).

136. B. Nilsson, 'Loshultskuppen,' in *Terra Scaniae*, educational website of the Skåne Regional Council (Skåne Regional Council). <http://www.ts.skane.se/> accessed 11 December 2012.

137. A. Åberg, *Snapphanarna*. (LTs förlag, Stockholm, 1952).

138. A. Zamoyski, *1812: Napoleon's Fatal March on Moscow*. (Harper Perennial, New York, 2004).

139. P. LeCouteur, J. Burreson, *Napoleon's Buttons*. (Jeremy P. Tarcher, New York, 2004).

140. R. Chang, *General Chemistry, 10th ed.* (McGraw-Hill, New York, 2010).

141. R. Petrucci, W. S. Harwood, J. D. Madura, *General Chemistry: Principles and Modern Applications*. (Pearson/Prentice Hall, Upper Saddle River, 2007).

142. L. J. Vionnet de Maringoné, *Campagnes de Russie & de Saxe (1812–1813): Souvenirs d'un Ex-commandant des Grenadiers de la Vieille-garde*. (E. Dubois, 1899 (original from the University of California, digitized 10 November 2007)).

143. J. Emsley, *Nature's Building Blocks: An A–Z Guide to the Elements*. (Oxford University Press, Oxford, 2001).

144. A number of webpages dedicated to military history that list in detail the composition of the uniforms of different regiments: <http://l.brenet.free.fr/grognards.htm> <http://pagesperso-orange.fr/eliedufaure1824–1865/militaire.htm> <http://www.tajan.com/pdf/2005/5580.pdf> <http://www.7cuirassiers.be/index.php?option=com_content&task=view&id=25&Itemid=34> accessed 11 December 2012.

145. M. Signoli, Y. Ardagna, P. Adalian, W. Devriendt, L. Lalys, C. Rigeade, T. Vette, A. Kuncevicius, J. Poskiene, A. Barkus, Z. Palubeckaite, A. Garmus, V. Pugaciauskas, R. Jankauskas, O. Dutour, 'Discovery of a Mass Grave of Napoleonic Period in Lithuania (1812, Vilnius),' *Comptes Rendus Palevol* 3, 219 (2004).

146. L. Tolstoy, *War and Peace*. (Oxford University Press, Oxford, 2010 (first published in Russian 1869)).

147. A. Zamoyski, Personal Communication, email correspondance (2008).

148. J. D. Dana, E. S. Dana, 'Item Notes: v. 177,' *The American Journal of Science* (1909).

149. E. Cohen, 'The Allotropy of Metals (A Lecture Delivered before the Faraday Society, Tuesday, June 13, 1911),' *Transactions of the Faraday Society* (1911).

150. C. Fritsche, 'Ueber eigenthumlich modificirtes Zinn,' *Berichte der Deutschen Chemischen Gesellschaft* 2, 112 (1869).

151. R. Baudin, O. Fraisse, 'Tout la lumière sur l'ètain,' *Bulletin de l'Union des Physiciens* 865, 961 (2004).

152. *Simple Steps to Protect Your Family from Lead Hazards.* (United States Environmental Protection Agency, United States Consumer Product Safety Commission, United States Department of Housing and Urban Development, EPA747-K-99-001, Washington DC, 2003).

153. M. V. Pollio, 'Book VIII,' in *The Ten Books on Architecture.* (Adamant Media Corporation, Chestnut Hill).

154. S. Mays, A. Ogden, J. Montgomery, S. Vincent, W. Battersby, G. M. Taylor, 'New Light on the Personal Identification of a Skeleton of a Member of Sir John Franklin's Last Expedition to the Arctic, 1845,' *Journal of Archaeological Science* 38, 1571 (2011).

155. W. Battersby, 'Identification of the Probable Source of the Lead Poisoning Observed in Members of the Franklin Expedition,' *Journal of the Hakluyt Society*, September (2008).

156. A. Keenleyside, X. Song, D. R. Chettle, C. E. Webber, 'The Lead Content of Human Bones from the 1845 Franklin Expedition,' *Journal of Archaeological Science* 23, 461 (1996).

157. W. J. Mills, *Exploring Polar Frontiers: A Historical Encyclopedia, Vol. 1.* (ABC-CLIO, Santa Barbara, 2003).

158. T. I. Lidsky, J. S. Schneider, 'Lead Neurotoxicity in Children: Basic Mechanisms and Clinical Correlates,' *Brain* 126, 5 (2003).

159. W. Kovarik, 'Ethyl-leaded Gasoline: How a Classic Occupational Disease became an International Public Health Disaster,' *International Journal of Occupational and Environmental Health* 11, 384 (2005).

160. W. Kovarik, *Charles F. Kettering and the 1921 Discovery of Tetraethyl Lead in the Context of Technological Alternatives.* (Originally presented to the Society of Automotive Engineers Fuels & Lubricants Conference, Baltmore, MD, 1994; revised in 1999). <http://www4.hmc.edu:8001/Chemistry/Pb/resources/Kovarik.pdf> accessed 12 June 2012.

161. D. Seyferth, 'The Rise and Fall of Tetraethyl-lead: 1. Discovery and Slow Development in European Universities, 1853–1920 (vol. 22, pp. 2346, 2003),' *Organometallics* 23, 1,172 (2004).

162. W. Kovarik, 'Ph.D. Thesis,' University of Maryland (1993).

163. D. Seyferth, 'The Rise and Fall of Tetraethyl-lead: 2,' *Organometallics* 22, 5,154 (2003).

164. A. Hamilton, P. Reznikoff, G. M. Burnham, 'Tetra-ethyl Lead,' *Journal of the American Medical Association* 84, 1481 (1925).

165. *Alice Hamilton Awards for Occupational Safety and Health.* (US National Institute of Occupational Safety and Health). <http://www.cdc.gov/niosh/awards/hamilton/> accessed 11 December 2012.

166. J. J. Otten, J. P. Hellwig, L. D. Meyers, eds, *Dietary Reference Intakes: The Essential Guide to Nutrient Requirements.* (The National Academies Press, Washington, 2006).

167. J. B. Vincent, 'Chromium: Celebrating 50 Years as an Essential Element?' *Dalton Transactions* 39, 3787 (2010).

168. J. B. Vincent, S. T. Love, 'The Need for Combined Inorganic, Biochemical, and Nutritional Studies of Chromium(III),' *Chemistry & Biodiversity* 9, 1923 (2012).

169. *The Transportation of Natural Gas.* (Natural Gas Supply Association, US, 2011). <http://www.naturalgas.org/naturalgas/transport.Asp> accessed 11 December 2012.

170. 'Toxic award?' *Science* 310, 229 (2005).

171. G. Kolata, 'A Hit Movie is Rated "F" in Science,' *New York Times*, 11 April 2000.

172. P. Bracchi, B. McMahon, 'She was the Single Mother who Claimed Her Town was Poisoned by its Water Supply...but Was Erin Brockovich Wrong?' *Daily Mail*, 2 January 2011.

173. J. P. Jacobs, 'Utilities Gird for New Regs as EPA Studies Toxicity of Hex Chromium,' *New York Times*, 28 April 2011.

174. *Hexavalent Chromium.* (U.S. Department of Labor, Occupational Safety and Health Administration). <http://www.osha.gov/Publications/OSHA-3373-hexavalent-chromium.pdf> accessed 11 December 2012.

175. C. Skrzycki, 'OSHA Slow to Issue Standards, Critics Charge,' *The Washington Post*, 9 November 2004.

176. K. Nakamuro, K. Yoshikawa, Y. Sayato, H. Kurata, 'Comparative Studies of Chromosomal Aberration and Mutagenicity of Trivalent and Hexavalent Chromium,' *Mutation Research* 58, 175–81 (1978).

177. B. Hileman, 'Balancing Panels, Charges Have Been Made that New Appointments to Committees are Politically Motivated,' *Chemical & Engineering News*, no. 10, 37–9, 10 March 2003.

178. *Hinkley Compressor Station Chromium Contamination Cleanup.* (Lahontan Regional Water Quality Control Board, California Environmental

Protection Agency, 2013). <http://www.swrcb.ca.gov/rwqcb6/water_issues/projects/pge/index.shtml> accessed 31 March 2013.

179. G. Nordström, 'Elisabeth Taylor till sjukhus,' *Expressen*, 20 February 1997.

180. 'Roxette Star's Surgery "A Success"', *BBC News*, 4 October 2002. <http://news.bbc.co.uk/2/hi/entertainment/2299745.stm> accessed 12 June 2013.

181. *Material Safety Data Sheet, Potassium Nitrate*. (Fisher Scientific, Hampton, 2008).

182. *Material Safety Data Sheet, Gadolinium(III) Nitrate Hexahydrate, 99.9%*. (Fisher Scientific, Hampton, 2008).

183. *Information on Gadolinium-based Contrast Agents*. (U.S. Food and Drug Administration, Washington DC, 2010). <http://www.fda.gov/Drugs/DrugSafety/PostmarketDrugSafetyInformationforPatientsandProviders/ucm142882.htm> accessed 2 April 2013.

184. *Brain Tumor: Diagnosis* (American Society of Clinical Oncology). <http://www.cancer.net/patient/Cancer+Types/Brain+Tumor?sectionTitle=Diagnosis> accessed 11 December 2012.

185. A. Wiggin, 'The Truth Behind China's Rare Earths Embargo,' *Forbes*, 20 October 2010.

186. E. Helmore, 'China's Stranglehold on Rare Earth Metals "No Threat to US Security"', *The Guardian*, 31 October 2010.

187. *Tasman's PEA Study of Norra Karr Heavy Rare Earth and Zirconium Project Demonstrates Robust Economics and Long Mine Life*. News Release, 21 March (Tasman Metals Ltd, Vancouver, 2012).

188. L. Sjöstrand, 'Erik XIV:s sinnessjukdom ett resultat av arv och dåligt samvete?' *Läkartidningen* 103, 3647 (2006).

189. L. Ericson, *Johan III: en biografi*. (Historiska Media, Lund, 2004).

190. I. Andersson, 'Erik XIV,' in *Svenskt biografiskt lexikon*. (Riksarkivet, Stockholm, 1953), vol. Band 14, p. 282.

191. T. Yeatts, *Forensics: Solving the Crime*. (Oliver Press, Inc, Minneapolis, 2001).

192. J. Emsley, 'Whatever Happened to Arsenic?' *New Scientist* 19, 10 (1985).

193. S. H. Webster, 'The Development of the Marsh Test for Arsenic,' *Journal of Chemical Education* 24, 487–90 (1947).

194. D. Sayers, *Strong Poison*. (Victor Gollancz, London, 1930).

195. K.-A. Sempler, 'Revolver-Harry och Kalle Blomqvist,' *NyTeknik*, 29 September 2001.

196. A. Lindgren, *Second triology titel: Mästerdetektiven Blomkvist lever farligt*. (English translation: *Bill Bergson Lives Dangerously*, Rabén & Sjögren, Stockholm, 1951 (English ed. 1954)).

197. S. Larsson, *First triology title: Män som hatar kvinnor.* (English translation: *The Girl with the Dragon Tattoo*, Norstedts, Stockholm, 2005).

198. B. K. Mandal, K. T. Suzuki, 'Arsenic Round the World: A Review,' *Talanta* 58, 201 (2002).

199. D. Chakraborti, M. M. Rahman, K. Paul, U. K. Chowdhury, M. K. Sengupta, D. Lodh, C. R. Chanda, K. C. Saha, S. C. Mukherjee, 'Arsenic Calamity in the Indian Subcontinent: What Lessons Have Been Learned?' *Talanta* 58, 3 (2002).

200. J. Wennerberg, *Läkemedel som förändrat världen.* (Apotekarsocieteten, Stockholm, 2012).

201. D. Sayers, R. Eustace, *The Documents in the Case.* (Victor Gollancz, London, 1930).

202. W. S. Knowles, 'Nobel Lecture,' in *Les Prix Nobel: The Nobel Prizes 2001*, T. Frängsmyr, ed. (Nobel Foundation, Stockholm, 2002).

203. C. Elschenbroich, A. Salzer, *Organometallics: A Concise Introduction.* (VCH, Weinheim, 2nd ed., 1989).

204. The Michael J. Fox Foundation for Parkinson's Research. <https://http://www.michaeljfox.org/> accessed 5 November 2012.

205. *Fast Facts About Caltech History.* (California Institute of Technology, Pasadena, 2009). <http://archives.caltech.edu/about/fastfacts.html> accessed 8 April 2013.

206. C. Cussler, D. Cussler, *Arctic Drift.* (Putnam, New York, 2008).

207. I. McEwan, *Solar.* (Random House, New York, 2010).

208. J. Griggs, 'Ian McEwan: Mr Sunshine,' *New Scientist*, 30 March 2010. <http://www.newscientist.com/blogs/culturelab/2010/03/ian-mcewan-mr-sunshine.html> accessed 12 June 2013.

INDEX